Understanding Photosynthesis

Understanding Photosynthesis

Wilfred Clancy

SYRAWOOD
PUBLISHING HOUSE

New York

Published by Syrawood Publishing House,
750 Third Avenue, 9th Floor,
New York, NY 10017, USA
www.syrawoodpublishinghouse.com

Understanding Photosynthesis
Wilfred Clancy

International Standard Book Number: 978-1-64740-102-3 (Hardback)

Cataloging-in-Publication Data

Understanding photosynthesis / Wilfred Clancy.
    p. cm.
Includes bibliographical references and index.
ISBN 978-1-64740-102-3
1. Photosynthesis. 2. Photobiology. 3. Plants--Effect of light on.
4. Plants--Photorespiration. I. Clancy, Wilfred.
QK882 .U53 2022
572.46--dc23

Table of Contents

Preface

This book is a culmination of my many years of practice in this field. I attribute the success of this book to my support group. I would like to thank my parents who have showered me with unconditional love and support and my peers and professors for their constant guidance.

Photosynthesis is the life-sustaining process used by plants and other organisms that converts light energy into chemical energy. This energy is later released to fuel the activities of organisms. Photosynthesis is responsible for the production and maintenance of oxygen content in the Earth's atmosphere. During this process, the energy containing carbohydrate molecules such as sugars, are synthesized from carbon dioxide and water. All of the organic compounds and most of the energy required for life on earth are supplied by photosynthesis. It begins with the absorption of energy from light by proteins containing chlorophyll pigments. The organisms which can perform photosynthesis, such as plants and algae, are called photoautotrophs. The topics included in this book on photosynthesis are of utmost significance and bound to provide incredible insights to readers. It covers in detail some existent theories and innovative concepts revolving around this area of study. Those in search of information to further their knowledge will be greatly assisted by this book.

The details of chapters are provided below for a progressive learning:

Chapter – What is Photosynthesis?

Photosynthesis refers to the process used by plants and other organisms for the conversion of light energy into chemical energy. This energy can be released to fuel the activities of organisms. The topics elaborated in this chapter will help in gaining a better perspective about various aspects of photosynthesis.

Chapter – Fundamental Aspects of Photosynthesis

Some of the major concepts related to photosynthesis are photosynthetic efficiency, PI curve, photorespiration, crassulacean acid metabolism, non-photochemical quenching and photosynthetically active radiation. This chapter has been carefully written to provide an easy understanding of these key concepts of photosynthesis.

Chapter – Photosynthetic Pigments and Enzymes

The pigments which are used to capture the light energy which is essential for photosynthesis are known as photosynthetic pigments. Some of the major photosynthetic pigments and enzymes are chlorophyll, bacteriochlorophyll, allophycocyanin, pheophytin and phycobilin. This chapter has been carefully written to provide an easy understanding of the varied facets of these photosynthetic pigments and enzymes.

Chapter – Light Reactions in Photosynthesis

The reactions which constitute the first stage of photosynthesis where light energy is converted into chemical energy are known as light-dependent reactions. The chapter closely examines the major types of light reactions in photosynthesis such as cyclic photophosphorylation and non-cyclic photophosphorylation.

Chapter – Phototrophic Bacteria

The group of bacteria which derive energy from the sun in order to grow are known as phototrophic bacteria. A few examples of such bacteria are heliobacteria, chlorobium, chlorobium chlorochromatii, chloroflexus aurantiacus, green sulfur bacteria, rhodobacter sphaeroides, purple bacteria and cyanobacteria. This chapter has been carefully written to provide an easy understanding of these types of phototrophic bacteria

Wilfred Clancy

1

What is Photosynthesis?

Photosynthesis refers to the process used by plants and other organisms for the conversion of light energy into chemical energy. This energy can be released to fuel the activities of organisms. The topics elaborated in this chapter will help in gaining a better perspective about various aspects of photosynthesis.

Photosynthesis is the process by which green plants and certain other organisms transform light energy into chemical energy. During photosynthesis in green plants, light energy is captured and used to convert water, carbon dioxide, and minerals into oxygen and energy-rich organic compounds.

It would be impossible to overestimate the importance of photosynthesis in the maintenance of life on Earth. If photosynthesis ceased, there would soon be little food or other organic matter on Earth. Most organisms would disappear, and in time Earth's atmosphere would become nearly devoid of gaseous oxygen. The only organisms able to exist under such conditions would be the chemosynthetic bacteria, which can utilize the chemical energy of certain inorganic compounds and thus are not dependent on the conversion of light energy.

Energy produced by photosynthesis carried out by plants millions of years ago is responsible for the fossil fuels (i.e., coal, oil, and gas) that power industrial society. In past ages, green plants and small organisms that fed on plants increased faster than they were consumed, and their remains were deposited in Earth's crust by sedimentation and other geological processes. There, protected from oxidation, these organic remains were slowly converted to fossil fuels. These fuels not only provide much of the energy used in factories, homes, and transportation but also serve as the raw material for plastics and other synthetic products. Unfortunately, modern civilization is using up in a few centuries the excess of photosynthetic production accumulated over millions of years. Consequently, the carbon dioxide that has been removed from the air to make carbohydrates in photosynthesis over millions of years is being returned at an incredibly rapid rate. The carbon dioxide concentration in Earth's atmosphere is rising the fastest it ever has in Earth's history, and this phenomenon is expected to have major implications on Earth's climate.

Requirements for food, materials, and energy in a world where human population is rapidly growing have created a need to increase both the amount of photosynthesis and the efficiency of converting photosynthetic output into products useful to people. One response to those needs—the so-called Green Revolution, begun in the mid-20th century—achieved enormous improvements in agricultural yield through the use of chemical fertilizers, pest and plant-disease control, plant breeding, and mechanized tilling, harvesting, and crop processing. This effort limited severe famines to a few areas of the world despite rapid population growth, but it did not eliminate widespread malnutrition. Moreover, beginning in the early 1990s, the rate at which yields of major crops increased began to decline. This was especially true for rice in Asia. Rising costs associated with sustaining high rates of agricultural production, which required ever-increasing inputs of fertilizers and pesticides and constant development of new plant varieties, also became problematic for farmers in many countries.

A second agricultural revolution, based on plant genetic engineering, was forecast to lead to increases in plant productivity and thereby partially alleviate malnutrition. Since the 1970s, molecular biologists have possessed the means to alter a plant's genetic material (deoxyribonucleic acid, or DNA) with the aim of achieving improvements in disease and drought resistance, product yield and quality, frost hardiness, and other desirable properties. However, such traits are inherently complex, and the process of making changes to crop plants through genetic engineering has turned out to be more complicated than anticipated. In the future such genetic engineering may result in improvements in the process of photosynthesis, but by the first decades of the 21st century, it had yet to demonstrate that it could dramatically increase crop yields.

Another intriguing area in the study of photosynthesis has been the discovery that certain animals are able to convert light energy into chemical energy. The emerald green sea slug (*Elysia chlorotica*), for example, acquires genes and chloroplasts from *Vaucheria litorea*, an alga it consumes, giving it a limited ability to produce chlorophyll. When enough chloroplasts are assimilated, the slug may forgo the ingestion of food. The pea aphid (*Acyrthosiphon pisum*) can harness light to manufacture the energy-rich compound adenosine triphosphate (ATP); this ability has been linked to the aphid's manufacture of carotenoid pigments.

Development of the Idea

The study of photosynthesis began in 1771 with observations made by the English clergyman and scientist Joseph Priestley. Priestley had burned a candle in a closed container until the air within the container could no longer support combustion. He then placed a sprig of mint plant in the container and discovered that after several days the mint had produced some substance (later recognized as oxygen) that enabled the confined air to again support combustion. In 1779 the Dutch physician Jan Ingenhousz expanded upon Priestley's work, showing that the plant had to be exposed to light if the combustible substance (i.e., oxygen) was to be restored. He also demonstrated that this process required the presence of the green tissues of the plant.

In 1782 it was demonstrated that the combustion-supporting gas (oxygen) was formed at the expense of another gas, or "fixed air," which had been identified the year before as carbon dioxide. Gas-exchange experiments in 1804 showed that the gain in weight of a plant grown in a carefully weighed pot resulted from the uptake of carbon, which came entirely from absorbed carbon dioxide, and water taken up by plant roots; the balance is oxygen, released back to the atmosphere. Almost half a century passed before the concept of chemical energy had developed sufficiently to permit the discovery (in 1845) that light energy from the sun is stored as chemical energy in products formed during photosynthesis.

Overall Reaction of Photosynthesis

In chemical terms, photosynthesis is a light-energized oxidation–reduction process. (Oxidation refers to the removal of electrons from a molecule; reduction refers to the gain of electrons by a molecule.) In plant photosynthesis, the energy of light is used to drive the oxidation of water (H_2O), producing oxygen gas (O_2), hydrogen ions (H^+), and electrons. Most of the removed electrons and hydrogen ions ultimately are transferred to carbon dioxide (CO_2), which is reduced to organic products. Other electrons and hydrogen ions are used to reduce nitrateand sulfate to amino and sulfhydryl groups in amino acids, which are the building blocks of proteins. In most green cells, carbohydrates—especially starch and the sugar sucrose—are the major direct organic products of photosynthesis. The overall reaction in which carbohydrates—represented by the general formula (CH_2O)—are formed during plant photosynthesis can be indicated by the following equation:

$$CO_2 + 2\,H_2O \xrightarrow[\text{green plant}]{\text{light}} (CH_2O) + O_2 + H_2O$$

This equation is merely a summary statement, for the process of photosynthesis actually involves numerous reactions catalyzed by enzymes (organic catalysts). These reactions occur in two stages: the "light" stage, consisting of photochemical (i.e., light-capturing) reactions; and the "dark" stage, comprising chemical reactions controlled by enzymes. During the first stage, the energy of light is absorbed and used to drive a series of electron transfers, resulting in the synthesis of ATP and the electron-donor-reduced nicotine adenine dinucleotide phosphate(NADPH). During the dark stage, the ATP and NADPH formed in the light-capturing reactions are used to reduce carbon dioxide to organic carbon compounds. This assimilation of inorganic carbon into organic compounds is called carbon fixation.

During the 20th century, comparisons between photosynthetic processes in green plants and in certain photosynthetic sulfur bacteria provided important information about the photosynthetic mechanism. Sulfur bacteria use hydrogen sulfide (H_2S) as a source of hydrogen atoms and produce sulfur instead of oxygen during photosynthesis. The overall reaction is:

$$CO_2 + 2H_2S \xrightarrow[\substack{\text{sulfur}\\ \text{bacteria}}]{\text{light}} (CH_2O) + S_2 + H_2O$$

In the 1930s Dutch biologist Cornelis van Niel recognized that the utilization of carbon dioxide to form organic compounds was similar in the two types of photosynthetic organisms. Suggesting that differences existed in the light-dependent stage and in the nature of the compounds used as a source of hydrogen atoms, he proposed that hydrogen was transferred from hydrogen sulfide (in bacteria) or water (in green plants) to an unknown acceptor (called A), which was reduced to H_2A. During the dark reactions, which are similar in both bacteria and green plants, the reduced acceptor (H_2A) reacted with carbon dioxide (CO_2) to form carbohydrate (CH_2O) and to oxidize the unknown acceptor to A. This putative reaction can be represented as:

$$CO_2 + 2H_2A \xrightarrow{\text{light}} (CH_2O) + 2A + H_2O$$

Van Niel's proposal was important because the popular (but incorrect) theory had been that oxygen was removed from carbon dioxide (rather than hydrogen from water, releasing oxygen) and that carbon then combined with water to form carbohydrate (rather than the hydrogen from water combining with CO_2 to form CH_2O).

By 1940 chemists were using heavy isotopes to follow the reactions of photosynthesis. Water marked with an isotope of oxygen (^{18}O) was used in early experiments. Plants that photosynthesized in the presence of water containing $H_2{}^{18}O$ produced oxygen gas containing ^{18}O; those that photosynthesized in the presence of normal water produced normal oxygen gas. These results provided definitive support for van Niel's theory that the oxygen gas produced during photosynthesis is derived from water.

Basic Products of Photosynthesis

As has been stated, carbohydrates are the most-important direct organic product of photosynthesis in the majority of green plants. The formation of a simple carbohydrate, glucose, is indicated by a chemical equation:

$$6CO_2 + 12H_2O \xrightarrow[\text{green plants}]{\text{light}} C_6H_{12}O_6 + 6O_2 + 6H_2O$$
$$\substack{\text{carbon}\\\text{dioxide}} \qquad \substack{\text{water}} \qquad\qquad\qquad \substack{\text{glucose}} \qquad \substack{\text{oxygen}} \qquad \substack{\text{water}}$$

Little free glucose is produced in plants; instead, glucose units are linked to form starch or are joined with fructose, another sugar, to form sucrose.

Not only carbohydrates, as was once thought, but also amino acids, proteins, lipids (or fats), pigments, and other organic components of green tissues are synthesized during photosynthesis. Minerals supply the elements (e.g., nitrogen, N; phosphorus, P; sulfur, S) required to form these compounds. Chemical bonds are broken between oxygen (O) and carbon (C), hydrogen (H), nitrogen, and sulfur, and new bonds are formed in products that include gaseous oxygen (O_2) and organic compounds. More energy is required to break the bonds between oxygen and other elements (e.g., in water, nitrate, and sulfate) than is released when new bonds form in the products. This difference in bond energy accounts for a large part of the light energy stored as chemical energy in the organic products formed during photosynthesis. Additional energy is stored in making complex molecules from simple ones.

Evolution of the Process

Although life and the quality of the atmosphere today depend on photosynthesis, it is likely that green plants evolved long after the first living cells. When Earth was young, electrical storms and solar radiation probably provided the energy for the synthesis of complex molecules from abundant simpler ones, such as water, ammonia, and methane. The first living cells probably evolved from these complex molecules. For example, the accidental joining (condensation) of the amino acid glycine and the fatty acidacetate may have formed complex organic molecules known as porphyrins. These molecules, in turn, may have evolved further into coloured molecules called pigments—e.g., chlorophyllsof green plants, bacteriochlorophyll of photosynthetic bacteria, hemin (the red pigment of blood), and cytochromes, a group of pigment molecules essential in both photosynthesis and cellular respiration.

Primitive coloured cells then had to evolve mechanisms for using the light energy absorbed by their pigments. At first, the energy may have been used immediately to initiate reactions useful to the cell. As the process for utilization of light energy continued to evolve, however, a larger part of the absorbed light energy probably was stored as chemical energy, to be used to maintain life. Green plants, with their ability to use light energy to convert carbon dioxide and water to carbohydrates and oxygen, are the culmination of this evolutionary process.

The first oxygenic (oxygen-producing) cells probably were the blue-green algae(cyanobacteria), which appeared about two billion to three billion years ago. These microscopic organisms are believed to have greatly increased the oxygen content of the atmosphere, making possible the development of aerobic (oxygen-using) organisms. Cyanophytes are prokaryotic cells; that is, they contain no distinct membrane-enclosed subcellular particles (organelles), such as nuclei and chloroplasts. Green plants, by contrast, are composed of eukaryotic cells, in which the photosynthetic apparatus is contained within membrane-bound chloroplasts. The complete genome sequences of cyanobacteria and higher plants provide evidence that the first photosynthetic eukaryotes were likely the red algae that developed when nonphotosynthetic eukaryotic cells engulfed cyanobacteria. Within the host cells, these cyanobacteria evolved into chloroplasts.

There are a number of photosynthetic bacteria that are not oxygenic (e.g., the sulfur bacteriapreviously discussed). The evolutionary pathway that led to these bacteria diverged from the one that resulted in oxygenic organisms. In addition to the absence of oxygen production, nonoxygenic photosynthesis differs from oxygenic photosynthesis in two other ways: light of longer wavelengths is absorbed and used by pigments called bacteriochlorophylls, and reduced compounds other than water (such as hydrogen sulfide or organic molecules) provide the electrons needed for the reduction of carbon dioxide.

Factors that Influence the Rate of Photosynthesis

The rate of photosynthesis is defined in terms of the rate of oxygen production either per unit mass (or area) of green plant tissues or per unit weight of total chlorophyll. The amount of light, the carbon dioxide supply, temperature, water supply, and the availability of minerals are the most important environmental factors that affect the rate of photosynthesis in land plants. The rate of photosynthesis is also determined by the plant species and its physiological state—e.g., its health, its maturity, and whether it is in flower.

Light Intensity and Temperature

As has been mentioned, the complex mechanism of photosynthesis includes a photochemical, or light-harvesting, stage and an enzymatic, or carbon-assimilating, stage that involves chemical reactions. These stages can be distinguished by studying the rates of photosynthesis at various degrees of light saturation (i.e., intensity) and at different temperatures. Over a range of moderate temperatures and at low to medium light intensities (relative to the normal range of the plant species), the rate of photosynthesis increases as the intensity increases and is relatively independent of temperature. As the light intensity increases to higher levels, however, the rate becomes saturated; light "saturation" is achieved at a specific light intensity, dependent on species and growing conditions. In the light-dependent range before saturation, therefore, the rate of photosynthesis is determined by the rates of photochemical steps. At high light intensities, some of the chemical reactions of the dark stage become rate-limiting. In many land plants, a process called photorespiration occurs, and its influence upon photosynthesis increases with rising temperatures. More specifically, photorespiration competes with photosynthesis and limits further increases in the rate of photosynthesis, especially if the supply of water is limited.

Carbon Dioxide

Included among the rate-limiting steps of the dark stage of photosynthesis are the chemical reactions by which organic compounds are formed by using carbon dioxide as a carbon source. The rates of these reactions can be increased somewhat by increasing the carbon dioxide concentration. Since the middle of the 19th century, the level of carbon dioxide in the atmosphere has been rising because of the extensive combustion of fossil fuels, cement production, and land-use changes associated with deforestation. The atmospheric level of carbon dioxide climbed from about 0.028 percent in 1860 to 0.032 percent by 1958 (when improved measurements began) and to 0.040 percent by 2016. This increase in carbon dioxide directly increases plant photosynthesis, but the size of the increase depends on the species and physiological condition of the plant. Furthermore, most scientists maintain that increasing levels of atmospheric carbon dioxide affect climate, increasing global temperatures and changing rainfall patterns. Such changes will also affect photosynthesis rates.

Water

For land plants, water availability can function as a limiting factor in photosynthesis and plant growth. Besides the requirement for a small amount of water in the photosynthetic reaction itself, large amounts of water are transpired from the leaves; that is, water evaporates from the leaves to the atmosphere via the stomata. Stomata are small openings through the leaf epidermis, or outer skin; they permit the entry of carbon dioxide but inevitably also allow the exit of water vapour. The stomata open and close according to the physiological needs of the leaf. In hot and arid climates the stomata may close to conserve water, but this closure limits the entry of carbon dioxide and hence the rate of photosynthesis. The decreased transpirationmeans there is less cooling of the leaves and hence leaf temperatures rise. The decreased carbon dioxide concentration inside the leaves and the increased leaf temperatures favour the wasteful process of photorespiration. If the level of carbon dioxide in the atmosphere increases, more carbon dioxide could enter through a smaller opening of the stomata, so more photosynthesis could occur with a given supply of water.

Minerals

Several minerals are required for healthy plant growth and for maximum rates of photosynthesis. Nitrogen, sulfate, phosphate, iron, magnesium, calcium, and potassium are required in substantial amounts for the synthesis of amino acids, proteins, coenzymes, deoxyribonucleic acid (DNA) and ribonucleic acid (RNA), chlorophyll and other pigments, and other essential plant constituents. Smaller amounts of such elements as manganese, copper, and chloride are required in photosynthesis. Some other trace elements are needed for various nonphotosynthetic functions in plants.

Internal Factors

Each plant species is adapted to a range of environmental factors. Within this normal range of conditions, complex regulatory mechanisms in the plant's cells adjust the activities of enzymes (i.e., organic catalysts). These adjustments maintain a balance in the overall photosynthetic process and control it in accordance with the needs of the whole plant. With a given plant species, for example, doubling the carbon dioxide level might cause a temporary increase of nearly twofold in the rate of photosynthesis; a few hours or days later, however, the rate might fall to the original level because photosynthesis produced more sucrose than the rest of the plant could use. By contrast, another plant species provided with such carbon dioxide enrichment might be able to use more sucrose, because it had more carbon-demanding organs, and would continue to photosynthesize and to grow faster throughout most of its life cycle.

Energy Efficiency of Photosynthesis

The energy efficiency of photosynthesis is the ratio of the energy stored to the energy of lightabsorbed. The chemical energy stored is the difference between that contained in gaseous oxygen and organic compound products and the energy of water, carbon dioxide, and other reactants. The amount of energy stored can only be estimated because many products are formed, and these vary with the plant species and environmental conditions. If the equation for glucose formation given earlier is used to approximate the actual storage process, the production of one mole (i.e., 6.02×10^{23} molecules; abbreviated N) of oxygen and one-sixth mole of glucose results in the storage of about 117 kilocalories (kcal) of chemical energy. This amount must then be compared with the energy of light absorbed to produce one mole of oxygen in order to calculate the efficiency of photosynthesis.

Light can be described as a wave of particles known as photons; these are units of energy, or light quanta. The quantity N photons is called an einstein. The energy of light varies inversely with the length of the photon waves; that is, the shorter the wavelength, the greater the energy content. The energy (e) of a photon is given by the equation $e = hc/\lambda$, where c is the velocity of light, h is Planck's constant, and λ is the light wavelength. The energy (E) of an einstein is $E = Ne = Nhc/\lambda = 28,600/\lambda$, when E is in kilocalories and λ is given in nanometres (nm; 1 nm = 10^{-9} metres). An einstein of red light with a wavelength of 680 nm has an energy of about 42 kcal. Blue light has a shorter wavelength and therefore more energy than red light. Regardless of whether the light is blue or red, however, the same number of einsteins are required for photosynthesis per mole of oxygen formed. The part of the solar spectrum used by plants has an estimated mean wavelength of 570 nm; therefore, the energy of light used during photosynthesis is approximately 28,600/570, or 50 kcal per einstein.

In order to compute the amount of light energy involved in photosynthesis, one other value is needed: the number of einsteins absorbed per mole of oxygen evolved. This is called the quantum requirement. The minimum quantum requirement for photosynthesis under optimal conditions is about nine. Thus, the energy used is 9 × 50, or 450 kcal per mole of oxygen evolved. Therefore, the estimated maximum energy efficiency of photosynthesis is the energy stored per mole of oxygen evolved, 117 kcal, divided by 450—that is, 117/450, or 26 percent.

The actual percentage of solar energy stored by plants is much less than the maximum energy efficiency of photosynthesis. An agricultural crop in which the biomass (total dry weight) stores as much as 1 percent of total solar energy received on an annual areawide basis is exceptional, although a few cases of higher yields (perhaps as much as 3.5 percent in sugarcane) have been reported. There are several reasons for this difference between the predicted maximum efficiency of photosynthesis and the actual energy stored in biomass. First, more than half of the incident sunlight is composed of wavelengths too long to be absorbed, and some of the remainder is reflected or lost to the leaves. Consequently, plants can at best absorb only about 34 percent of the incident sunlight. Second, plants must carry out a variety of physiological processes in such nonphotosynthetic tissues as roots and stems; these processes, as well as cellular respiration in all parts of the plant, use up stored energy. Third, rates of photosynthesis in bright sunlight sometimes exceed the needs of the plants, resulting in the formation of excess sugars and starch. When this happens, the regulatory mechanisms of the plant slow down the process of photosynthesis, allowing more absorbed sunlight to go unused. Fourth, in many plants, energy is wasted by the process of photorespiration. Finally, the growing season may last only a few months of the year; sunlight received during other seasons is not used. Furthermore, it should be noted that if only agricultural products (e.g., seeds, fruits, and tubers, rather than total biomass) are considered as the end product of the energy-conversion process of photosynthesis, the efficiency falls even further.

Chloroplasts, the Photosynthetic Units of Green Plants

The process of plant photosynthesis takes place entirely within the chloroplasts. Detailed studies of the role of these organelles date from the work of British biochemist Robert Hill. About 1940 Hill discovered that green particles obtained from broken cells could produce oxygen from water in the presence of light and a chemical compound, such as ferric oxalate, able to serve as an electron acceptor. This process is known as the Hill reaction. During the 1950s Daniel Arnon and other American biochemists prepared plant cell fragments in which not only the Hill reaction but also the synthesis of the energy-storage compound ATP occurred. In addition, the coenzyme NADP was used as the final acceptor of electrons, replacing the nonphysiological electron acceptors used by Hill. His procedures were refined further so that small individual pieces of isolated chloroplast membranes, or lamellae, could perform the Hill reaction. These small pieces of lamellae were then fragmented into pieces so small that they performed only the light reactions of the photosynthetic process. It is now possible also to isolate the entire chloroplast so that it can carry out the complete process of photosynthesis, from light absorption, oxygen formation, and the reduction of carbon dioxide to the formation of glucose and other products.

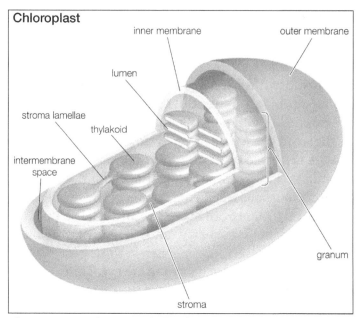

Chloroplast structure: The internal (thylakoid) membrane vesicles are organized into stacks, which reside in a matrix known as the stroma. All the chlorophyll in the chloroplast is contained in the membranes of the thylakoid vesicles.

Process of Photosynthesis: The Light Reactions

Light Absorption and Energy Transfer

The light energy absorbed by a chlorophyll molecule excites some electrons within the structure of the molecule to higher energy levels, or excited states. Light of shorter wavelength (such as blue) has more energy than light of longer wavelength (such as red), so absorption of blue light creates an excited state of higher energy. A molecule raised to this higher energy state quickly gives up the "extra" energy as heat and falls to its lowest excited state. This lowest excited state is similar to that of a molecule that has just absorbed the longest wavelength light capable of exciting it. In the case of chlorophyll a, this lowest excited state corresponds to that of a molecule that has absorbed red light of about 680 nm.

The return of a chlorophyll a molecule from its lowest excited state to its original low-energy state (ground state) requires the release of the extra energy of the excited state. This can occur in one of several ways. In photosynthesis, most of this energy is conserved as chemical energyby the transfer of an electron from a special chlorophyll a molecule (P_{680} or P_{700}) to an electron acceptor. When this electron transfer is blocked by inhibitors, such as the herbicide dichlorophenylmethylurea (DCMU), or by low temperature, the energy can be released as red light. Such reemission of light is called fluorescence. The examination of fluorescence from photosynthetic material in which electron transfer has been blocked has proved to be a valuable tool for scientists studying the light reactions.

The Pathway of Electrons

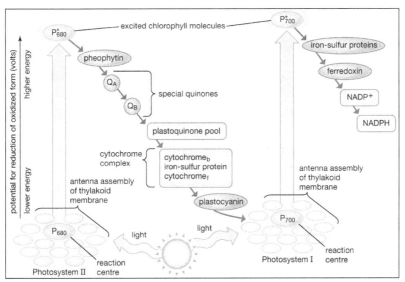

Flow of electrons during the light reaction stage of photosynthesis arrows pointing upward represent light reactions that increase the chemical potential; arrows slanting downward represent flow of electrons via carriers in the membrane.

The general features of a widely accepted mechanism for photoelectron transfer, in which two light reactions (light reaction I and light reaction II) occur during the transfer of electrons from water to carbon dioxide, were proposed by Robert Hill and Fay Bendall in 1960. This mechanism is based on the relative potential (in volts) of various cofactors of the electron-transfer chain to be oxidized or reduced. Molecules that in their oxidized form have the strongest affinity for electrons (i.e., are strong oxidizing agents) have a low relative potential. In contrast, molecules that in their oxidized form are difficult to reduce have a high relative potential once they have accepted electrons. The molecules with a low relative potential are considered to be strong oxidizing agents, and those with a high relative potential are considered to be strong reducing agents.

In diagrams that describe the light reaction stage of photosynthesis, the actual photochemical steps are typically represented by two vertical arrows. These arrows signify that the special pigments P_{680} and P_{700} receive light energy from the light-harvesting chlorophyll-protein molecules and are raised in energy from their ground state to excited states. In their excited state, these pigments are extremely strong reducing agents that quickly transfer electrons to the first acceptor. These first acceptors also are strong reducing agents and rapidly pass electrons to more stable carriers. In light reaction II, the first acceptor may be pheophytin, which is a molecule similar to chlorophyll that also has a strong reducing potential and quickly transfers electrons to the next acceptor. Special quinones are next in the series. These molecules are similar to plastoquinone; they receive electrons from pheophytin and pass them to the intermediate electron carriers, which include the plastoquinone pool and the cytochromes b and f associated in a complex with an iron-sulfur protein.

In light reaction I, electrons are passed on to iron-sulfur proteins in the lamellar membrane, after which the electrons flow to ferredoxin, a small water-soluble iron-sulfur protein. When $NADP^+$ and a suitable enzyme are present, two ferredoxin molecules, carrying one electron each, transfer two electrons to $NADP^+$, which picks up a proton (i.e., a hydrogen ion) and becomes NADPH.

Each time a P_{680} or P_{700} molecule gives up an electron, it returns to its ground (unexcited) state, but with a positive charge due to the loss of the electron. These positively charged ions are extremely strong oxidizing agents that remove an electron from a suitable donor. The P_{680}^+ of light reaction II is capable of taking electrons from water in the presence of appropriate catalysts. There is good evidence that two or more manganese atoms complexed with protein are involved in this catalysis, taking four electrons from two water molecules (with release of four hydrogen ions). The manganese-protein complex gives up these electrons one at a time via an unidentified carrier to P_{680}^+, reducing it to P_{680}. When manganese is selectively removed by chemical treatment, the thylakoids lose the capacity to oxidize water, but all other parts of the electron pathway remain intact.

In light reaction I, P_{700}^+ recovers electrons from plastocyanin, which in turn receives them from intermediate carriers, including the plastoquinone pool and cytochrome b and cytochrome f molecules. The pool of intermediate carriers may receive electrons from water via light reaction II and the quinones. Transfer of electrons from water to ferredoxin via the two light reactions and intermediate carriers is called noncyclic electron flow. Alternatively, electrons may be transferred only by light reaction I, in which case they are recycled from ferredoxin back to the intermediate carriers. This process is called cyclic electron flow.

Evidence of Two Light Reactions

Many lines of evidence support the concept of electron flow via two light reactions. An early study by American biochemist Robert Emerson employed the algae *Chlorella*, which was illuminated with red light alone, with blue light alone, and with red and blue light at the same time. Oxygen evolution was measured in each case. It was substantial with blue light alone but not with red light alone. With both red and blue light together, the amount of oxygen evolved far exceeded the sum of that seen with blue and red light alone. These experimental data pointed to the existence of two types of light reactions that, when operating in tandem, would yield the highest rate of oxygen evolution. It is now known that light reaction I can use light of a slightly longer wavelength than red ($\lambda = 680$ nm), while light reaction II requires light with a wavelength of 680 nm or shorter.

Since those early studies, the two light reactions have been separated in many ways, including separation of the membrane particles in which each reaction occurs. As discussed previously, lamellae can be disrupted mechanically into fragments that absorb light energy and break the bonds of water molecules (i.e., oxidize water) to produce oxygen, hydrogen ions, and electrons. These electrons can be transferred to ferredoxin, the final electron acceptor of the light stage. No transfer of electrons from water to ferredoxin occurs if the herbicide DCMU is present. The subsequent addition of certain reduced dyes (i.e., electron donors) restores the light reductionof NADP$^+$ but without oxygen production, suggesting that light reaction I but not light reaction II is functioning. It is now known that DCMU blocks the transfer of electrons between the first quinone and the plastoquinone pool in light reaction II.

When treated with certain detergents, lamellae can be broken down into smaller particles capable of carrying out single light reactions. One type of particle can absorb light energy, oxidize water, and produce oxygen (light reaction II), but a special dye molecule must be supplied to accept the electrons. In the presence of electron donors, such as a reduced dye, a second type of lamellar particle can absorb light and transfer electrons from the electron donor to ferredoxin (light reaction I).

Photosystems I and II

The structural and photochemical properties of the minimum particles capable of performing light reactions I and II have received much study. Treatment of lamellar fragments with neutral detergents releases these particles, designated photosystem I and photosystem II, respectively. Subsequent harsher treatment (with charged detergents) and separation of the individual polypeptides with electrophoretic techniques have helped identify the components of the photosystems. Each photosystem consists of a light-harvesting complex and a core complex. Each core complex contains a reaction centre with the pigment (either P_{700} or P_{680}) that can be photochemically oxidized, together with electron acceptors and electron donors. In addition, the core complex has some 40 to 60 chlorophyll molecules bound to proteins. In addition to the light absorbed by the chlorophyll molecules in the core complex, the reaction centres receive a major part of their excitation from the pigments of the light-harvesting complex.

Quantum Requirements

The quantum requirements of the individual light reactions of photosynthesis are defined as the number of light photons absorbed for the transfer of one electron. The quantum requirement for each light reaction has been found to be approximately one photon. The total number of quanta required, therefore, to transfer the four electrons that result in the formation of one molecule of oxygen via the two light reactions should be four times two, or eight. It appears, however, that additional light is absorbed and used to form ATP by a cyclic photophosphorylation pathway. (The cyclic photophosphorylation pathway is an ATP-forming process in which the excited electron returns to the reaction centre.) The actual quantum requirement, therefore, probably is 9 to 10.

Process of Photosynthesis: The Conversion of Light Energy to ATP

The electron transfers of the light reactions provide the energy for the synthesis of two compounds vital to the dark reactions: NADPH and ATP. The previous section explained how noncyclic electron flow results in the reduction of NADP⁺ to NADPH. In this section, the synthesis of the energy-rich compound ATP is described.

ATP is formed by the addition of a phosphate group to a molecule of adenosine diphosphate(ADP)—or to state it in chemical terms, by the phosphorylation of ADP. This reaction requires a substantial input of energy, much of which is captured in the bond that links the added phosphate group to ADP. Because light energy powers this reaction in the chloroplasts, the production of ATP during photosynthesis is referred to as photophosphorylation, as opposed to oxidative phosphorylation in the electron-transport chain in the mitochondrion.

Unlike the production of NADPH, the photophosphorylation of ADP occurs in conjunction with both cyclic and noncyclic electron flow. In fact, researchers speculate that the sole purpose of cyclic electron flow may be for photophosphorylation, since this process involves no net transfer of electrons to reducing agents. The relative amounts of cyclic and noncyclic flow may be adjusted in accordance with changing physiological needs for ATP and reduced ferredoxin and NADPH in chloroplasts. In contrast to electron transfer in light reactions I and II, which can occur in membrane fragments, intact thylakoids are required for efficient photophosphorylation. This requirement stems from the special nature of the mechanism linking photophosphorylation to electron flow in the lamellae.

The theory relating the formation of ATP to electron flow in the membranes of both chloroplasts and mitochondria (the organelles responsible for ATP formation during cellular respiration) was first proposed by English biochemist Peter Dennis Mitchell, who received the 1978 Nobel Prize for Chemistry. This chemiosmotic theory has been somewhat modified to fit later experimental facts. The general features are now widely accepted. A central feature is the formation of a hydrogen ion (proton) concentration gradient and an electrical charge across intact lamellae. The potential energy stored by the proton gradient and electrical charge is then used to drive the energetically unfavourable conversion of ADP and inorganic phosphate (P_i) to ATP and water.

The manganese-protein complex associated with light reaction II is exposed to the interior of the thylakoid. Consequently, the oxidation of water during light reaction II leads to release of hydrogen ions (protons) into the inner thylakoid space. Furthermore, it is likely that photoreaction II entails the transfer of electrons across the lamella toward its outer face, so that when plastoquinone molecules are reduced, they can receive protons from the outside of the thylakoid. When these reduced plastoquinone molecules are oxidized, giving up electrons to the cytochrome-iron-sulfur complex, protons are released inside the thylakoid. Because the lamella is imperme-able to them, the release of protons inside the thylakoid by oxidation of both water and plastoquinone leads to a higher concentration of protons inside the thylakoid than outside it. In other words, a proton gradient is established across the lamella. Since protons are positively charged, the movement of protons across the thylakoid lamella during both light reactions results in the establishment of an electrical charge across the lamella.

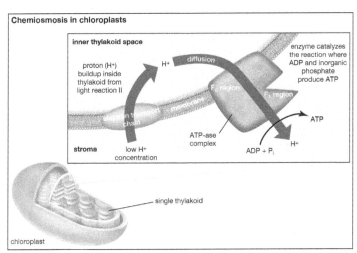

Chemiosmosis in chloroplasts that results in the donation of a
proton for the production of adenosine triphosphate (ATP) in plants.

An enzyme complex located partly in and on the lamellae catalyzes the reaction in which ATP is formed from ADP and inorganic phosphate. The reverse of this reaction is catalyzed by an enzyme called ATP-ase; hence, the enzyme complex is sometimes called an ATP-ase complex. It is also called the coupling factor. It consists of hydrophil-ic polypeptides (F_1), which project from the outer surface of the lamellae, and hydro-phobic polypeptides (F_0), which are embedded inside the lamellae. F_0 forms a channel that permits protons to flow through the lamellar membrane to F_1. The enzymes in F_1 then catalyze ATP formation, using both the proton supply and the lamellar transmem-brane charge.

In summary, the use of light energy for ATP formation occurs indirectly: a proton gra-dient and electrical charge—built up in or across the lamellae as a consequence of elec-tron flow in the light reactions—provide the energy to drive the synthesis of ATP from ADP and P_i.

Process of Photosynthesis: Carbon Fixation and Reduction

The assimilation of carbon into organic compounds is the result of a complex series of enzymatically regulated chemical reactions—the dark reactions. This term is something of a misnomer, for these reactions can take place in either light or darkness. Further-more, some of the enzymes involved in the so-called dark reactions become inactive in prolonged darkness; however, they are activated when the leaves that contain them are exposed to light.

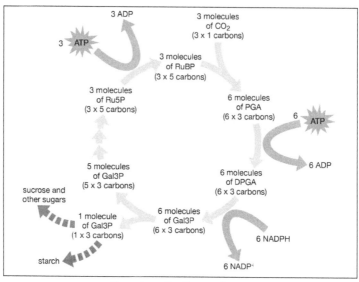

C_3 carbon fixation pathway.

Pathway of carbon dioxide fixation and reduction in photosynthesis, the reductive pentose phosphate cycle. The diagram represents one complete turn of the cycle, with the net production of one molecule of Gal3P. The nine molecules of ATP and six molecules of NADPH come from the light reactions.

Elucidation of the Carbon Pathway

Radioactive isotopes of carbon (^{14}C) and phosphorus (^{32}P) have been valuable in identifying the intermediate compounds formed during carbon assimilation. A photosynthesizing plant does not strongly discriminate between the most abundant natural carbon isotope (^{12}C) and ^{14}C. During photosynthesis in the presence of $^{14}CO_2$, the compounds formed become labeled with the radioisotope. During very short exposures, only the first intermediates in the carbon-fixing pathway become labeled. Early investigations showed that some radioactive products were formed even when the light was turned off and the $^{14}CO_2$ was added just afterward in the dark, confirming the nature of the carbon fixation as a "dark" reaction.

American biochemist Melvin Calvin, a Nobel Prize recipient for his work on the carbon-reduction cycle, allowed green plants to photosynthesize in the presence of radioactive carbon dioxide for a few seconds under various experimental conditions. Products that became labeled with radioactive carbon during Calvin's experiments included a three-carbon compound called 3-phosphoglycerate (abbreviated PGA), sugar phosphates, amino acids, sucrose, and carboxylic acids. When photosynthesis was stopped after two seconds, the principal radioactive product was PGA, which therefore was identified as the first stable compound formed during carbon dioxide fixation in green plants. PGA is a three-carbon compound, and the mode of photosynthesis is thus referred to as C_3. In the two other known pathways, C_4 and CAM (crassulacean acid metabolism), the C_3 pathway follows the fixation of CO_2 into oxaloacetate, a four-carbon acid, and its reduction to malate. PGA is formed from 2-carboxy-3-keto-D-arabinitol 1,5-bisphosphate, which is a highly unstable six-carbon compound formed from the carboxylation of ribulose-1,5-bisphosphate, a five-carbon compound.

Further studies with ^{14}C as well as with inorganic phosphate labeled with ^{32}P led to the mapping of the carbon fixation and reduction pathway called the reductive pentose phosphate (RPP) cycle, or the Calvin-Benson cycle. An additional pathway for carbon transport in certain plants was later discovered in other laboratories. All the steps in these pathways can be carried out in the laboratory by isolated enzymes in the dark. Several steps require the ATP or NADPH generated by the light reactions. In addition, some of the enzymes are fully active only when conditions simulate those in green cells exposed to light. In living plants, these enzymes are active during photosynthesis but not in the dark.

The Calvin-benson Cycle

The Calvin-Benson cycle, in which carbon is fixed, reduced, and utilized, involves the formation of intermediate sugar phosphates in a cyclic sequence. One complete cycle incorporates three molecules of carbon dioxide and produces one molecule of the three-carbon compound glyceraldehyde-3-phosphate (Gal3P). This three-carbon sugar phosphate usually is either exported from the chloroplasts or converted to starch inside the chloroplast.

ATP and NADPH formed during the light reactions are utilized for key steps in this pathway and provide the energy and reducing equivalents (i.e., electrons) to drive the sequence in the direction shown. For each molecule of carbon dioxide that is fixed, two molecules of NADPH and three molecules of ATP from the light reactions are required. The overall reaction can be represented as follows:

$$9 \text{ATP} + 6 \text{NADPH} + 3 CO_2 \rightarrow \text{Gal} 3 P + 6 \text{NADPH}^+ + 9 \text{ADP} + 8 P_i$$

The cycle is composed of four stages: (1) carboxylation, (2) reduction, (3) isomerization/condensation/dismutation, and (4) phosphorylation.

Carboxylation

The initial incorporation of carbon dioxide, which is catalyzed by the enzyme ribulose 1,5-bisphosphate carboxylase (Rubisco), proceeds by the addition of carbon dioxide to the five-carbon compound ribulose 1,5-bisphosphate (RuBP) and the splitting of the resulting six-carbon compound into two molecules of PGA. This reaction occurs three times during each complete turn of the cycle; thus, six molecules of PGA are produced.

Reduction

The six molecules of PGA are first phosphorylated with ATP by the enzyme PGA-kinase, yielding six molecules of 1,3-diphosphoglycerate (DPGA). These molecules are subsequently reduced with NADPH and the enzyme glyceraldehyde-3-phosphate dehydrogenase to give six molecules of Gal3P. These reactions are the reverse of two steps of the process glycolysis in cellular respiration.

Isomerization/Condensation/Dismutation

For each complete Calvin-Benson cycle, one of the Gal3P molecules, with its three carbon atoms, is the net product and may be transferred out of the chloroplast or converted to starch inside the chloroplast. For the cycle to regenerate, the other five Gal3P molecules (with a total of 15 carbon atoms) must be converted back to three molecules of five-carbon RuBP. The conversion of Gal3P to RuBP begins with a complex series of enzymatically regulated reactions that lead to the synthesis of the five-carbon compound ribulose-5-phosphate (Ru5P).

Phosphorylation

The three molecules of Ru5P are converted to the carboxylation substrate, RuBP, by the enzyme phosphoribulokinase, using ATP. This reaction, shown below, completes the cycle:

$$3\,Ru\,5\,P + 3_{ATP} \rightarrow 3\,Ru\,BP + 3_{ADP}$$

Regulation of the Cycle

Photosynthesis cannot occur at night, but the respiratory process of glycolysis—which uses some of the same reactions as the Calvin-Benson cycle, except in the reverse—does take place. Thus, some steps in this cycle would be wasteful if allowed to occur in the dark, because they would counteract the reactions of glycolysis. For this reason, some enzymes of the Calvin-Benson cycle are "turned off" (i.e., become inactive) in the dark.

Even in the presence of light, changes in physiological conditions frequently necessitate adjustments in the relative rates of reactions of the Calvin-Benson cycle, so that enzymes for some reactions change in their catalytic activity. These alterations in enzyme activity typically are brought about by changes in levels of such chloroplast components as reduced ferredoxin, acids, and soluble components (e.g., P_i and magnesium ions).

Products of Carbon Reduction

The most important use of Gal3P is its export from the chloroplasts to the cytosol of green cells, where it is used for biosynthesis of products needed by the plant. In land plants, a principal product is sucrose, which is translocated from the green cells of the leaves to other parts of the plant. Other key products include the carbon skeletons of certain primary amino acids, such as alanine, glutamate, and aspartate. To complete the synthesis of these compounds, amino groups are added to the appropriate carbon skeletons made from Gal3P. Sulfur amino acids such as cysteine are formed by adding sulfhydryl groups and amino groups. Other biosynthesis pathways lead from Gal3P to lipids, pigments, and most of the constituents of green cells.

Starch synthesis and accumulation in the chloroplasts occur particularly when photosynthetic carbon fixation exceeds the needs of the plant. Under such circumstances, sugar phosphates accumulate in the cytosol, binding cytosolic P_i. The export of Gal3P from the chloroplasts is tied to a one-for-one exchange of P_i for Gal3P, so less cytosolic P_i results in decreased export of Gal3P and decreased P_i in the chloroplast. These changes trigger alterations in the activities of regulated enzymes, leading in turn to increased starch synthesis. This starch can be broken down at night and used as a source of reduced carbon and energy for the physiological needs of the plant. Too much starch in the chloroplasts leads to diminished rates of photosynthesis, however. In addition, high levels of sugars in the cytosol lead to the suppression of the normal activities of the genes involved in photosynthesis. Thus, under what would seem to be the ideal photosynthetic conditions of a bright warm day, many plants in fact have-slower-than expected rates of photosynthesis.

Photorespiration

Under conditions of high light intensity, hot weather, and water limitation, the productivity of the Calvin-Benson cycle is limited in many plants by the occurrence of photorespiration. This process converts sugar phosphates back to carbon dioxide; it is initiated by the oxygenation of RuBP (i.e., the combination of gaseous oxygen [O_2] with RuBP). This oxygenation reaction yields only one molecule of PGA and one molecule of a two-carbon acid, phosphoglycolate, which is subsequently converted in part to carbon dioxide. The reaction of oxygen with RuBP is in direct competition with the carboxylation reaction (CO_2 + RuBP) that initiates the Calvin-Benson cycle and is, in fact, catalyzed by the same protein, ribulose 1,5-bisphosphate carboxylase. The relative concentrations of oxygen and carbon dioxide within the chloroplasts as well as leaf temperature determine whether oxygenation or carboxylation is favoured. The concentration of oxygen inside the chloroplasts may be higher than atmospheric (20 percent) because of photosynthetic oxygen evolution, whereas the internal carbon dioxide concentration may be lower than atmospheric (0.039 percent) because of photosynthetic uptake. Any increase in the internal carbon dioxide pressure tends to help the carboxylation reaction compete more effectively with oxygenation.

Carbon Fixation in C_4 Plants

Certain plants—including the important crops sugarcane and corn (maize), as well as other diverse species that are thought to have expanded their geographic ranges into tropical areas—have developed a special mechanism of carbon fixation that largely prevents photorespiration. The leaves of these plants have special anatomy and biochemistry. In particular, photosynthetic functions are divided between mesophyll and bundle-sheath leafcells. The carbon-fixation pathway begins in the mesophyll cells, where carbon dioxide is converted into bicarbonate, which is then added to the three-carbon acid phosphoenolpyruvate (PEP) by an enzyme called phosphoenolpyruvate carboxylase. The product of this reaction is the four-carbon acid oxaloacetate, which is reduced to malate, another four-carbon acid, in one form of the C_4 pathway. Malate then is transported to bundle-sheath cells, which are located near the vascular system of the leaf. There, malate enters the chloroplasts and is oxidized and decarboxylated (i.e., loses CO_2) by malic enzyme. This yields high concentrations of carbon dioxide, which is fed into the Calvin-Benson cycle of the bundle sheath cells, and pyruvate, a three-carbon acid that is translocated back to the mesophyll cells. In the mesophyll chloroplasts, the enzyme pyruvate orthophosphate dikinase (PPDK) uses ATPand P_i to convert pyruvate back to PEP, completing the C_4 cycle. There are several variations of this pathway in different species. For example, the amino acids aspartate and alanine can substitute for malate and pyruvate in some species.

The C_4 pathway acts as a mechanism to build up high concentrations of carbon dioxide in the chloroplasts of the bundle sheath cells. The resulting higher level of internal carbon dioxide in these chloroplasts serves to increase the ratio of carboxylation to oxygenation, thus minimizing photorespiration. Although the plant must expend extra energy to drive this mechanism, the energy loss is more than compensated by the near elimination of photorespiration under conditions where it would otherwise occur. Sugarcane and certain other plants that employ this pathway have the highest annual yields of biomass of all species. In cool climates, where photorespiration is insignificant, C_4 plants are rare. Carbon dioxide is also used efficiently in carbohydrate synthesis in the bundle sheath.

PEP carboxylase, which is located in the mesophyll cells, is an essential enzyme in C_4 plants. In hot and dry environments, carbon dioxide concentrations inside the leaf fall when the plant closes or partially closes its stomata to reduce water loss from the leaves. Under these conditions, photorespiration is likely to occur in plants that use Rubisco as the primary carboxylating enzyme, since Rubisco adds oxygen to RuBP when carbon dioxide concentrations are low. PEP carboxylase, however, does not use oxygen as a substrate, and it has a greater affinity for carbon dioxide than Rubisco does. Thus, it has the ability to fix carbon dioxide in reduced carbon dioxide conditions, such as when the stomata on the leaves are only partially open. As a consequence, at similar rates of photosynthesis, C_4 plants lose less water when compared with C_3 plants. This explains why C_4 plants are favoured in dry and warm environments.

Carbon Fixation Via Crassulacean Acid Metabolism (CAM)

Prickly pear cactus (*Opuntia*).

In addition to C_3 and C_4 species, there are many succulent plants that make use of a third photosynthetic pathway: crassulacean acid metabolism (CAM). This pathway is named after the Crassulaceae, a family in which many species display this type of metabolism, but it also occurs commonly in other families, such as the Cactaceae, the Euphorbiaceae, the Orchidaceae, and the Bromeliaceae. CAM species number more than 20,000 and span 34 families. Almost all CAM plants are angiosperms; however, quillworts and ferns also use the CAM pathway. In addition, some scientists note that CAM might be used by *Welwitschia*, a gymnosperm. CAM plants are often characterized by their succulence, but this quality is not pronounced in epiphytes that use the CAM pathway.

CAM plants are known for their capacity to fix carbon dioxide at night, using PEP carboxylase as the primary carboxylating enzyme and the accumulation of malate (which is made by the enzyme malate dehydrogenase) in the large vacuoles of their cells. Deacidification occurs during the day, when carbon dioxide is released from malate and fixed in the Calvin-Benson cycle, using Rubisco. During daylight hours, the stomata are closed to prevent water loss. The stomata are open at night when the air is cooler and more humid, and this setting allows the leaves of the plant to assimilate carbon dioxide. Since their stomata are closed during the day, CAM plants require considerably less water than both C_3 and C_4 plants that fix the same amount of carbon dioxide in photosynthesis.

The productivity of most CAM plants is fairly low, however. This is not an inherent trait of CAM species, because some cultivated CAM plants (e.g., *Agave mapisaga* and *A. salmiana*) can achieve a high aboveground productivity. In fact, some cultivated species that are irrigated, fertilized, and carefully pruned are highly productive. For example, prickly pear (*Opuntia ficus-indica*) and its thornless variety, *O. amyclea*, produce 4.6 kg per square metre (0.9 pound per square foot) of new growth per year. Such productivity is among the highest of any plant species. Thus, the rates of photosynthesis of CAM plants may be as high as those of C_3 plants, if morphologically similar plants adapted to the similar habitats are compared.

The unusual capacity of CAM plants to fix carbon dioxide into organic acids in the dark, causing nocturnal acidification, with deacidification occurring during the day, has been known to science since the 19th century. (There is evidence, however, that the Romans noticed the difference between the morning acid taste of some of the house plants they cultivated.) On the other hand, the C_4 pathway was discovered during the middle of the 20th century. A full appreciation of CAM as a photosynthetic pathway was greatly stimulated by analogies with C_4 species.

Differences in Carbon Fixation Pathways

A comparison of the differences between the various carbon pathways is provided in the table.

Differences in the major carbon-fixation pathways in plants					
pathway	carbon-assimilation process	first stable intermediate product	stomate activity	photorespiration	plant types using this pathway
C3	Calvin-Benson cycle only	phosphoglycerate (PGA), a three-carbon acid	open during the day, closed at night	not suppressed	plants living incolder, wetter environments characterized by low-to-medium light intensities
C4	adds CO_2 to phosphoenolpyruvate (PEP) to form oxaloacetate first; the Calvin-Benson cycle follows	oxaloacetate, a four-carbon acid, which is later reduced to malate	open during the day, closed at night	suppressed	plants living in warmer, drier environments characterized by high light intensity
CAM*	adds CO_2 to phosphoenolpyruvate (PEP) to form oxaloacetate first; the Calvin-Benson cycle follows	oxaloacetate, a four-carbon acid, which is later reduced to malate and stored in vacuoles	open at night,closed during the day	suppressed	succulents (members of Crassulaceae), which occur in warmer, drier environments characterized by high light intensity

pathway	carbon-assimilation process	first stable intermediate product	stomate activity	photorespiration	plant types using this pathway
*Crassulacean acid metabolism.					

Molecular Biology of Photosynthesis

Oxygenic photosynthesis occurs in a certain type of prokaryotic cells called cyanobacteria and eukaryotic plant cells (algae and higher plants). In eukaryotic plant cells, which contain chloroplasts and a nucleus, the genetic information needed for the reproduction of the photosynthetic apparatus is contained partly in the chloroplast chromosome and partly in chromosomes of the nucleus. For example, the carboxylation enzyme ribulose 1,5-bisphosphate carboxylase is a large protein molecule comprising a complex of eight large polypeptide subunits and eight small polypeptide subunits. The gene for the large subunits is located in the chloroplast chromosome, whereas the gene for the small subunits is in the nucleus. Transcription of the DNA of the nuclear gene yields messenger RNA (mRNA) that encodes the information for the synthesis of the small polypeptides. During this synthesis, which occurs on the cytosolic ribosomes, some extra amino acid residues are added to form a recognition leader on the end of the polypeptide chain. This leader is recognized by special receptor sites on the outer chloroplast membrane; these receptor sites then allow the polypeptide to penetrate the membrane and enter the chloroplast. The leader is removed, and the small subunits combine with the large subunits, which have been synthesized on chloroplast ribosomes according to mRNA transcribed from the chloroplast DNA. The expression of nuclear genes that code for proteins needed in the chloroplasts appears to be under control of events in the chloroplasts in some cases; for example, the synthesis of some nuclear-encoded chloroplast enzymes may occur only when light is absorbed by chloroplasts.

2

Fundamental Aspects of Photosynthesis

Some of the major concepts related to photosynthesis are photosynthetic efficiency, PI curve, photorespiration, crassulacean acid metabolism, non-photochemical quenching and photosynthetically active radiation. This chapter has been carefully written to provide an easy understanding of these key concepts of photosynthesis.

PLASTID

Plastids are double-membrane organelle which are found in the cells of plants and algae. Plastids are responsible for manufacturing and storing of food. These often contain pigments that are used in photosynthesis and different types of pigments that can change the colour of the cell.

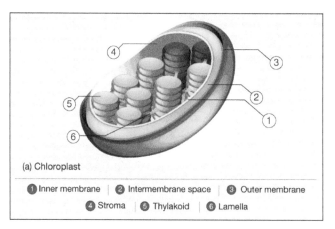

(a) Chloroplast

❶ Inner membrane   ❷ Intermembrane space   ❸ Outer membrane
❹ Stroma   ❺ Thylakoid   ❻ Lamella

Types of Plastids

There are different types of plastids with their specialized functions. Among which few are mainly classified based on the presence or absence of the Biological pigments and their stages of development:

- Chloroplasts

- Chromoplasts

- Gerontoplasts

- Leucoplasts

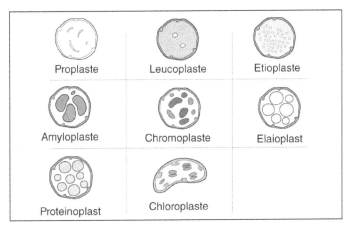

Chloroplasts

Chloroplasts are biconvex shaped, semi-porous, double membraned, permeable cell organelle found within the mesophyll of the plant cell. They are the sites for synthesis food by the process of photosynthesis.

Chromoplasts

Chromoplasts is the name given to an area for all the pigments to be kept and synthesized in the plant. These can be usually found in flowering plants, aging leaves and fruits. Chloroplasts convert into chromoplasts. Chromoplasts are carotenoid pigments that allow different colours that you see in leaves and fruits. The main reason for its structure and the colour for attracting pollinators.

Gerontoplasts

These are basically chloroplasts that go with the aging process. Geronoplasts refers to the chloroplasts of the leaves that helps the beginning to convert into different other organelles when the leaf is no longer using photosynthesis usually in an autumn month.

Leucoplasts

These are the non-pigmented organelles which are colourless. Leucoplasts are usually found in most of the non-photosynthetic parts of the plant like roots. They act as a storage sheds for starches, lipids, and proteins depending on the needs of the plants. They are mostly used for converting amino acids and fatty acids.

Leucoplasts are of three types:

- Amyloplasts – Amyloplasts are greatest among all three – Amyloplasts, proteinoplasts and elaioplasts and are easily charged with storing starch.

- Proteinoplasts – Proteinoplasts helps in storing the proteins that a plant needs and can be typically found in seeds.

- Elaioplasts -Elaioplasts helps in storing fats and oils that are needed by the plant.

Inheritance of Plastids

There are many plants which are inherited from the plastids from just a single parent. Angiosperms inherit plastids from the female gamete while there are many gymnosperms that inherit plastids from the male pollen. Algae inherit plastids from one parent only. The inheritance of the plastids DNA seems to be 100% uniparental. In hybridisation, the inheritance of plastid seems to be more erratic.

CHLOROPLAST

Chloroplasts are organelles that conduct photosynthesis, where the photosynthetic pigment chlorophyll captures the energy from sunlight, converts it, and stores it in the energy-storage molecules ATP and NADPH while freeing oxygen from water in plant and algal cells. They then use the ATP and NADPH to make organic molecules from carbon dioxide in a process known as the Calvin cycle. Chloroplasts carry out a number of other functions, including fatty acid synthesis, much amino acid synthesis, and the immune response in plants. The number of chloroplasts per cell varies from one, in unicellular algae, up to 100 in plants like *Arabidopsis* and wheat.

A chloroplast is a type of organelle known as a plastid, characterized by its two membranes and a high concentration of chlorophyll. Other plastid types, such as the leucoplast and the chromoplast, contain little chlorophyll and do not carry out photosynthesis.

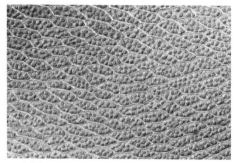

Chloroplasts visible in the cells of *Bryum capillare*, a type of moss.

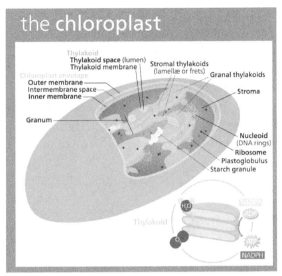

Structure of a typical higher-plant chloroplast.

Chloroplasts are highly dynamic—they circulate and are moved around within plant cells, and occasionally pinch in two to reproduce. Their behavior is strongly influenced by environmental factors like light color and intensity. Chloroplasts, like mitochondria, contain their own DNA, which is thought to be inherited from their ancestor—a photosynthetic cyanobacterium that was engulfed by an early eukaryotic cell. Chloroplasts cannot be made by the plant cell and must be inherited by each daughter cell during cell division.

With one exception (the amoeboid *Paulinella chromatophora*), all chloroplasts can probably be traced back to a single endosymbiotic event, when a cyanobacterium was engulfed by the eukaryote. Despite this, chloroplasts can be found in an extremely wide set of organisms, some not even directly related to each other—a consequence of many secondary and even tertiary endosymbiotic events.

Lineages and Evolution

Chloroplasts are one of many types of organelles in the plant cell. They are considered to have originated from cyanobacteria through endosymbiosis—when a eukaryotic cell engulfed a photosynthesizing cyanobacterium that became a permanent resident in the cell. Mitochondria are thought to have come from a similar event, where an aerobic prokaryote was engulfed. This origin of chloroplasts was first suggested by the Russian biologist Konstantin Mereschkowski in 1905 after Andreas Schimper observed in 1883 that chloroplasts closely resemble cyanobacteria. Chloroplasts are only found in plants, algae, and the amoeboid *Paulinella chromatophora*.

Cyanobacterial Ancestor

Cyanobacteria are considered the ancestors of chloroplasts. They are sometimes called blue-green algae even though they are prokaryotes. They are a diverse phylum of bacteria capable of carrying out photosynthesis, and are gram-negative, meaning that they have two cell membranes. Cyanobacteria also contain a peptidoglycan cell wall, which is thicker than in other gram-negative bacteria, and which is located between their two cell membranes. Like chloroplasts, they have thylakoids within. On the thylakoid membranes are photosynthetic pigments, including chlorophyll *a*. Phycobilins are also common cyanobacterial pigments, usually organized into hemispherical phycobilisomes attached to the outside of the thylakoid membranes (phycobilins are not shared with all chloroplasts though).

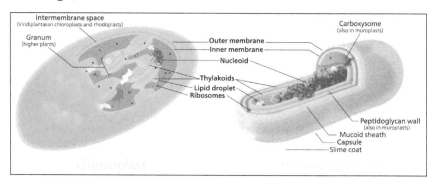

Both chloroplasts and cyanobacteria have a double membrane, DNA, ribosomes, and thylakoids. Both the chloroplast and cyanobacterium depicted are idealized versions (the chloroplast is that of a higher plant)—a lot of diversity exists among chloroplasts and cyanobacteria.

Primary Endosymbiosis

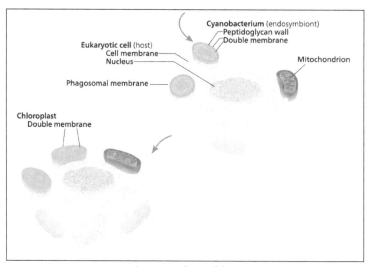

Primary endosymbiosis.

A eukaryote with mitochondria engulfed a cyanobacterium in an event of serial primary endosymbiosis, creating a lineage of cells with both organelles. It is important to note that the cyanobacterial endosymbiont already had a double membrane—the phagosomal vacuole-derived membrane was lost.

Somewhere around 1 to 2 billion years ago, a free-living cyanobacterium entered an early eukaryotic cell, either as food or as an internal parasite, but managed to escape the phagocytic vacuole it was contained in. The two innermost lipid-bilayer membranes that surround all chloroplasts correspond to the outer and inner membranes of the ancestral cyanobacterium's gram negative cell wall, and not the phagosomal membrane from the host, which was probably lost. The new cellular resident quickly became an advantage, providing food for the eukaryotic host, which allowed it to live within it. Over time, the cyanobacterium was assimilated, and many of its genes were lost or transferred to the nucleus of the host. From genomes that probably originally contained over 3000 genes only about 130 genes remain in the chloroplasts of contemporary plants. Some of its proteins were then synthesized in the cytoplasm of the host cell, and imported back into the chloroplast (formerly the cyanobacterium). Separately, somewhere around 500 million years ago, it happened again and led to the amoeboid *Paulinella chromatophora*.

This event is called *endosymbiosis*, or "cell living inside another cell with a mutual benefit for both". The external cell is commonly referred to as the *host* while the internal cell is called the *endosymbiont*.

Chloroplasts are believed to have arisen after mitochondria, since all eukaryotes contain mitochondria, but not all have chloroplasts. This is called *serial endosymbiosis*—an early eukaryote engulfing the mitochondrion ancestor, and some descendants of it then engulfing the chloroplast ancestor, creating a cell with both chloroplasts and mitochondria.

Whether or not primary chloroplasts came from a single endosymbiotic event, or many independent engulfments across various eukaryotic lineages, has long been debated. It is now generally held that organisms with primary chloroplasts share a single ancestor that took in a cyanobacterium 600–2000 million years ago. It has been proposed this bacterium was *Gloeomargarita lithophora*. The exception is the amoeboid *Paulinella chromatophora*, which descends from an ancestor that took in a *Prochlorococcus* cyanobacterium 90–500 million years ago.

These chloroplasts, which can be traced back directly to a cyanobacterial ancestor, are known as *primary plastids* ("*plastid*" in this context means almost the same thing as chloroplast). All primary chloroplasts belong to one of four chloroplast lineages—the glaucophyte chloroplast lineage, the amoeboid *Paulinella chromatophora* lineage, the rhodophyte (red algal) chloroplast lineage, or the chloroplastidan (green) chloroplast lineage. The rhodophyte and chloroplastidan lineages are the largest, with chloroplastidan (green) being the one that contains the land plants.

Glaucophyta

The alga *Cyanophora*, a glaucophyte, is thought to be one of the first organisms to contain a chloroplast. The glaucophyte chloroplast group is the smallest of the three primary chloroplast lineages, being found in only 13 species, and is thought to be the one that branched off the earliest. Glaucophytes have chloroplasts that retain a peptidoglycan wall between their double membranes, like their cyanobacterial parent. For this reason, glaucophyte chloroplasts are also known as 'muroplasts' (besides 'cyanoplasts' or 'cyanelles'). Glaucophyte chloroplasts also contain concentric unstacked thylakoids, which surround a carboxysome – an icosahedral structure that glaucophyte chloroplasts and cyanobacteria keep their carbon fixation enzyme RuBisCO in. The starch that they synthesize collects outside the chloroplast. Like cyanobacteria, glaucophyte and rhodophyte chloroplast thylakoids are studded with light collecting structures called phycobilisomes. For these reasons, glaucophyte chloroplasts are considered a primitive intermediate between cyanobacteria and the more evolved chloroplasts in red algae and plants.

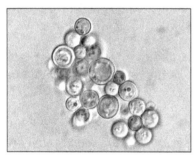 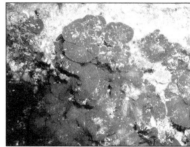 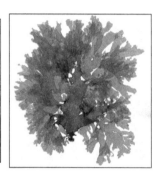

Diversity of red algae from left: *Cyanidium, Laurencia, Callophyllis laciniata*.
Red algal chloroplasts are characterized by phycobilin pigments
which often give them their reddish color.

Rhodophyceae (Red Algae)

The rhodophyte, or red algal chloroplast group is another large and diverse chloroplast lineage. Rhodophyte chloroplasts are also called *rhodoplasts*, literally "red chloroplasts".

Rhodoplasts have a double membrane with an intermembrane space and phycobilin pigments organized into phycobilisomes on the thylakoid membranes, preventing their thylakoids from stacking. Some contain pyrenoids. Rhodoplasts have chlorophyll *a* and phycobilins for photosynthetic pigments; the phycobilin phycoerytherin is responsible for giving many red algae their distinctive red color. However, since they also contain the blue-green chlorophyll *a* and other pigments, many are reddish to purple from the combination. The red phycoerytherin pigment is an adaptation to help red algae catch more sunlight in deep water—as such, some red algae that live in shallow water have less phycoerytherin in their rhodoplasts, and can appear more greenish. Rhodoplasts synthesize a form of starch called floridean starch, which collects into granules outside the rhodoplast, in the cytoplasm of the red alga.

Chloroplastida (Green Algae and Plants)

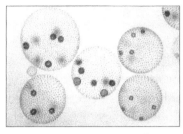 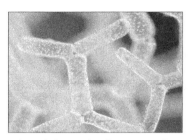 

Diversity of green algae from left: *Hydrodictyon*, *Volvox*, *Stigeoclonium*.
Green algal chloroplasts are characterized by their pigments chlorophyll
a and chlorophyll *b* which give them their green color.

The chloroplastidan chloroplasts, or green chloroplasts, are another large, highly diverse primary chloroplast lineage. Their host organisms are commonly known as the green algae and land plants. They differ from glaucophyte and red algal chloroplasts in that they have lost their phycobilisomes, and contain chlorophyll *b* instead. Most green chloroplasts are (obviously) green, though some aren't, like some forms of *Hæmatococcus pluvialis*, due to accessory pigments that override the chlorophylls' green colors. Chloroplastidan chloroplasts have lost the peptidoglycan wall between their double membrane, leaving an intermembrane space. Some plants seem to have kept the genes for the synthesis of the peptidoglycan layer, though they've been repurposed for use in chloroplast division instead.

Green algae and plants keep their starch *inside* their chloroplasts, and in plants and some algae, the chloroplast thylakoids are arranged in grana stacks. Some green algal chloroplasts contain a structure called a pyrenoid, which is functionally similar to the glaucophyte carboxysome in that it is where RuBisCO and CO_2 are concentrated in the chloroplast.

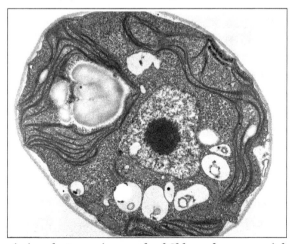

Transmission electron micrograph of *Chlamydomonas reinhardtii*,
a green alga that contains a pyrenoid surrounded by starch.

Helicosporidium is a genus of nonphotosynthetic parasitic green algae that is thought to contain a vestigial chloroplast. Genes from a chloroplast and nuclear genes indicating the presence of a chloroplast have been found in *Helicosporidium* even if nobody's seen the chloroplast itself.

Paulinella Chromatophora

While most chloroplasts originate from that first set of endosymbiotic events, *Paulinella chromatophora* is an exception that acquired a photosynthetic cyanobacterial endosymbiont more recently. It is not clear whether that symbiont is closely related to the ancestral chloroplast of other eukaryotes. Being in the early stages of endosymbiosis, *Paulinella chromatophora* can offer some insights into how chloroplasts evolved. *Paulinella* cells contain one or two sausage shaped blue-green photosynthesizing structures called chromatophores, descended from the cyanobacterium *Synechococcus*. Chromatophores cannot survive outside their host. Chromatophore DNA is about a million base pairs long, containing around 850 protein encoding genes—far less than the three million base pair *Synechococcus* genome, but much larger than the approximately 150,000 base pair genome of the more assimilated chloroplast. Chromatophores have transferred much less of their DNA to the nucleus of their host. About 0.3–0.8% of the nuclear DNA in *Paulinella* is from the chromatophore, compared with 11–14% from the chloroplast in plants.

Secondary and Tertiary Endosymbiosis

Many other organisms obtained chloroplasts from the primary chloroplast lineages through secondary endosymbiosis—engulfing a red or green alga that contained a chloroplast. These chloroplasts are known as secondary plastids.

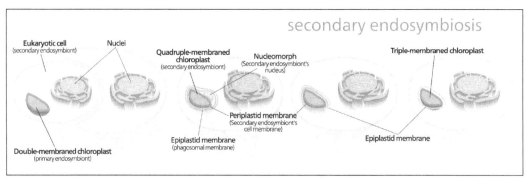

Secondary endosymbiosis consisted of a eukaryotic alga being engulfed by another
eukaryote, forming a chloroplast with three or four membranes.

While primary chloroplasts have a double membrane from their cyanobacterial ancestor, secondary chloroplasts have additional membranes outside of the original two, as a result of the secondary endosymbiotic event, when a nonphotosynthetic eukaryote engulfed a chloroplast-containing alga but failed to digest it—much like the cyanobacterium at the beginning of this story. The engulfed alga was broken down, leaving only its chloroplast, and sometimes its cell membrane and nucleus, forming a chloroplast with three or four membranes—the two cyanobacterial membranes, sometimes the eaten alga's cell membrane, and the phagosomal vacuole from the host's cell membrane.

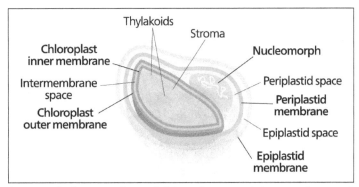

Diagram of a four membraned chloroplast containing a nucleomorph.

The genes in the phagocytosed eukaryote's nucleus are often transferred to the secondary host's nucleus. Cryptomonads and chlorarachniophytes retain the phagocytosed eukaryote's nucleus, an object called a nucleomorph, located between the second and third membranes of the chloroplast.

All secondary chloroplasts come from green and red algae—no secondary chloroplasts from glaucophytes have been observed, probably because glaucophytes are relatively rare in nature, making them less likely to have been taken up by another eukaryote.

Green Algal Derived Chloroplasts

Euglena, a euglenophyte, contains secondary chloroplasts from green algae.

Green algae have been taken up by the euglenids, chlorarachniophytes, a lineage of dinoflagellates, and possibly the ancestor of the CASH lineage (cryptomonads, alveolates, stramenopiles and haptophytes) in three or four separate engulfments. Many green algal derived chloroplasts contain pyrenoids, but unlike chloroplasts in their green algal ancestors, storage product collects in granules outside the chloroplast.

Euglenophytes

Euglenophytes are a group of common flagellated protists that contain chloroplasts derived from a green alga. Euglenophyte chloroplasts have three membranes—it is thought that the membrane of the primary endosymbiont was lost, leaving the cyanobacterial membranes, and the secondary host's phagosomal membrane. Euglenophyte chloroplasts have a pyrenoid and thylakoids stacked in groups of three. Photosynthetic product is stored in the form of paramylon, which is contained in membrane-bound granules in the cytoplasm of the euglenophyte.

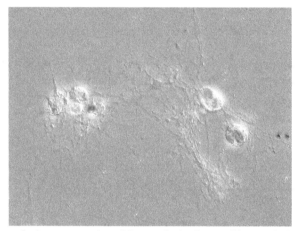

Chlorarachnion reptans is a chlorarachniophyte. Chlorarachniophytes replaced their original red algal endosymbiont with a green alga.

Chlorarachniophytes

Chlorarachniophytes are a rare group of organisms that also contain chloroplasts derived from green algae, though their story is more complicated than that of the euglenophytes. The ancestor of chlorarachniophytes is thought to have been a eukaryote with a *red* algal derived chloroplast. It is then thought to have lost its first red algal chloroplast, and later engulfed a green alga, giving it its second, green algal derived chloroplast.

Chlorarachniophyte chloroplasts are bounded by four membranes, except near the cell membrane, where the chloroplast membranes fuse into a double membrane. Their thylakoids are arranged in loose stacks of three. Chlorarachniophytes have a form of polysaccharide called chrysolaminarin, which they store in the cytoplasm, often collected around the chloroplast pyrenoid, which bulges into the cytoplasm.

Chlorarachniophyte chloroplasts are notable because the green alga they are derived from has not been completely broken down—its nucleus still persists as a nucleomorph found between the second and third chloroplast membranes—the periplastid space, which corresponds to the green alga's cytoplasm.

Prasinophyte-derived Dinophyte Chloroplast

Lepidodinium viride and its close relatives are dinophytes that lost their original peridinin chloroplast and replaced it with a green algal derived chloroplast (more specifically, a prasinophyte). *Lepidodinium* is the only dinophyte that has a chloroplast that's not from the rhodoplast lineage. The chloroplast is surrounded by two membranes and has no nucleomorph—all the nucleomorph genes have been transferred to the dinophyte nucleus. The endosymbiotic event that led to this chloroplast was serial secondary endosymbiosis rather than tertiary endosymbiosis—the endosymbiont was a green alga containing a primary chloroplast (making a secondary chloroplast).

Red Algal Derived Chloroplasts

Cryptophytes

Cryptophytes, or cryptomonads are a group of algae that contain a red-algal derived chloroplast. Cryptophyte chloroplasts contain a nucleomorph that superficially resembles that of the chlorarachniophytes. Cryptophyte chloroplasts have four membranes, the outermost of which is continuous with the rough endoplasmic reticulum. They synthesize ordinary starch, which is stored in granules found in the periplastid space—outside the original double membrane, in the place that corresponds to the red alga's cytoplasm. Inside cryptophyte chloroplasts is a pyrenoid and thylakoids in stacks of two.

Their chloroplasts do not have phycobilisomes, but they do have phycobilin pigments which they keep in their thylakoid space, rather than anchored on the outside of their thylakoid membranes.

Cryptophytes may have played a key role in the spreading of red algal based chloroplasts.

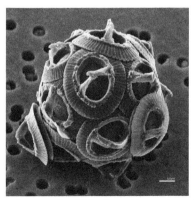

Scanning electron micrograph of *Gephyrocapsa oceanica*, a haptophyte.

Haptophytes

Haptophytes are similar and closely related to cryptophytes or heterokontophytes. Their chloroplasts lack a nucleomorph, their thylakoids are in stacks of three, and they synthesize chrysolaminarin sugar, which they store completely outside of the chloroplast, in the cytoplasm of the haptophyte.

Heterokontophytes (Stramenopiles)

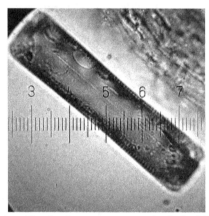

The photosynthetic pigments present in their chloroplasts give diatoms a greenish-brown color.

The heterokontophytes, also known as the stramenopiles, are a very large and diverse group of eukaryotes. The photoautotrophic lineage, Ochrophyta, including the diatoms and the brown algae, golden algae, and yellow-green algae, also contains red algal derived chloroplasts.

Heterokont chloroplasts are very similar to haptophyte chloroplasts, containing a pyrenoid, triplet thylakoids, and with some exceptions, having four layer plastidic envelope, the outermost epiplastid membrane connected to the endoplasmic reticulum. Like haptophytes, heterokontophytes store sugar in chrysolaminarin granules in the cytoplasm. Heterokontophyte chloroplasts contain chlorophyll a and with a few exceptions chlorophyll c, but also have carotenoids which give them their many colors.

Apicomplexans, Chromerids and Dinophytes

The alveolates are a major clade of unicellular eukaryotes of both autotrophic and heterotrophic members. The most notable shared characteristic is the presence of cortical (outer-region) alveoli (sacs). These are flattened vesicles (sacs) packed into a continuous layer just under the membrane and supporting it, typically forming a flexible pellicle (thin skin). In dinoflagellates they often form armor plates. Many members contain a red-algal derived plastid. One notable characteristic of this diverse group is the frequent loss of photosynthesis. However, a majority of these heterotrophs continue to process a non-photosynthetic plastid.

Apicomplexans

Apicomplexans are a group of alveolates. Like the helicosproidia, they're parasitic, and have a nonphotosynthetic chloroplast. They were once thought to be related to the helicosproidia, but it is now known that the helicosproida are green algae rather than part of the CASH lineage. The apicomplexans include *Plasmodium*, the malaria parasite. Many apicomplexans keep a vestigial red algal derived chloroplast called an apicoplast, which they inherited from their ancestors. Other apicomplexans like *Cryptosporidium* have lost the chloroplast completely. Apicomplexans store their energy in amylopectin granules that are located in their cytoplasm, even though they are nonphotosynthetic.

Apicoplasts have lost all photosynthetic function, and contain no photosynthetic pigments or true thylakoids. They are bounded by four membranes, but the membranes are not connected to the endoplasmic reticulum. The fact that apicomplexans still keep their nonphotosynthetic chloroplast around demonstrates how the chloroplast carries out important functions other than photosynthesis. Plant chloroplasts provide plant cells with many important things besides sugar, and apicoplasts are no different—they synthesize fatty acids, isopentenyl pyrophosphate, iron-sulfur clusters, and carry out part of the heme pathway. This makes the apicoplast an attractive target for drugs to cure apicomplexan-related diseases. The most important apicoplast function is isopentenyl pyrophosphate synthesis—in fact, apicomplexans die when something interferes with this apicoplast function, and when apicomplexans are grown in an isopentenyl pyrophosphate-rich medium, they dump the organelle.

Chromerids

The Chromerida is a newly discovered group of algae from Australian corals which comprises some close photosynthetic relatives of the apicomplexans. The first member, *Chromera velia*, was discovered and first isolated in 2001. The discovery of *Chromera velia* with similar structure to the apicomplexanss, provides an important link in the evolutionary history of the apicomplexans and dinophytes. Their plastids have four membranes, lack chlorophyll c and use the type II form of RuBisCO obtained from a horizontal transfer event.

Dinophytes

The dinoflagellates are yet another very large and diverse group of protists, around half of which are (at least partially) photosynthetic.

Most dinophyte chloroplasts are secondary red algal derived chloroplasts. Many other dinophytes have lost the chloroplast (becoming the nonphotosynthetic kind of dinoflagellate), or replaced it though *tertiary* endosymbiosis—the engulfment of another eukaryotic algae containing a red algal derived chloroplast. Others replaced their original chloroplast with a green algal derived one.

Most dinophyte chloroplasts contain form II RuBisCO, at least the photosynthetic pigments chlorophyll *a*, chlorophyll *c2*, *beta*-carotene, and at least one dinophyte-unique xanthophyll (peridinin, dinoxanthin, or diadinoxanthin), giving many a golden-brown color. All dinophytes store starch in their cytoplasm, and most have chloroplasts with thylakoids arranged in stacks of three.

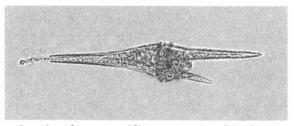

Ceratium furca, a peridinin-containing dinophyte.

The most common dinophyte chloroplast is the peridinin-type chloroplast, characterized by the carotenoid pigment peridinin in their chloroplasts, along with chlorophyll *a* and chlorophyll *c2*. Peridinin is not found in any other group of chloroplasts. The peridinin chloroplast is bounded by three membranes (occasionally two), having lost the red algal endosymbiont's original cell membrane. The outermost membrane is not connected to the endoplasmic reticulum. They contain a pyrenoid, and have triplet-stacked thylakoids. Starch is found outside the chloroplast. An important feature of these chloroplasts is that their chloroplast DNA is highly reduced and fragmented into many small circles. Most of the genome has migrated to the nucleus, and only critical photosynthesis-related genes remain in the chloroplast.

The peridinin chloroplast is thought to be the dinophytes' "original" chloroplast, which has been lost, reduced, replaced, or has company in several other dinophyte lineages.

Fucoxanthin-containing (Haptophyte-derived) Dinophyte Chloroplasts

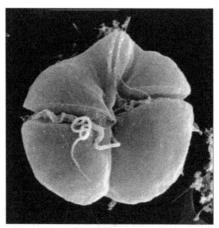

Karenia brevis is a fucoxanthin-containing dynophyte responsible for algal blooms called "red tides".

The fucoxanthin dinophyte lineages (including *Karlodinium* and *Karenia*) lost their original red algal derived chloroplast, and replaced it with a new chloroplast derived from a haptophyte endosymbiont. *Karlodinium* and *Karenia* probably took up different heterokontophytes. Because the haptophyte chloroplast has four membranes, tertiary endosymbiosis would be expected to create a six membraned chloroplast, adding the haptophyte's cell membrane and the dinophyte's phagosomal vacuole. However, the haptophyte was heavily reduced, stripped of a few membranes and its nucleus, leaving only its chloroplast (with its original double membrane), and possibly one or two additional membranes around it.

Fucoxanthin-containing chloroplasts are characterized by having the pigment fucoxanthin (actually 19′-hexanoyloxy-fucoxanthin and/or 19′-butanoyloxy-fucoxanthin) and no peridinin. Fucoxanthin is also found in haptophyte chloroplasts, providing evidence of ancestry.

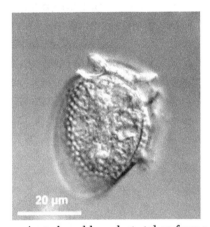

Dinophysis acuminata has chloroplasts taken from a cryptophyte.

Diatom-derived Dinophyte Chloroplasts

Some dinophytes, like *Kryptoperidinium* and *Durinskia* have a diatom (heterokontophyte) derived chloroplast. These chloroplasts are bounded by up to *five* membranes, (depending on whether you count the entire diatom endosymbiont as the chloroplast, or just the red algal derived chloroplast inside it). The diatom endosymbiont has been reduced relatively little—it still retains its original mitochondria, and has endoplasmic reticulum, ribosomes, a nucleus, and of course, red algal derived chloroplasts—practically a complete cell, all inside the host's endoplasmic reticulum lumen. However the diatom endosymbiont can't store its own food—its storage polysaccharide is found in granules in the dinophyte host's cytoplasm instead. The diatom endosymbiont's nucleus is present, but it probably can't be called a nucleomorph because it shows no sign of genome reduction, and might have even been *expanded*. Diatoms have been engulfed by dinoflagellates at least three times.

The diatom endosymbiont is bounded by a single membrane, inside it are chloroplasts with four membranes. Like the diatom endosymbiont's diatom ancestor, the chloroplasts have triplet thylakoids and pyrenoids.

In some of these genera, the diatom endosymbiont's chloroplasts aren't the only chloroplasts in the dinophyte. The original three-membraned peridinin chloroplast is still around, converted to an eyespot.

Kleptoplastidy

In some groups of mixotrophic protists, like some dinoflagellates (e.g. *Dinophysis*), chloroplasts are separated from a captured alga and used temporarily. These klepto chloroplasts may only have a lifetime of a few days and are then replaced.

Cryptophyte-derived Dinophyte Chloroplast

Members of the genus *Dinophysis* have a phycobilin-containing chloroplast taken from a cryptophyte. However, the cryptophyte is not an endosymbiont—only the chloroplast seems to have been taken, and the chloroplast has been stripped of its nucleomorph and outermost two membranes, leaving just a two-membraned chloroplast. Cryptophyte chloroplasts require their nucleomorph to maintain themselves, and *Dinophysis* species grown in cell culture alone cannot survive, so it is possible (but not confirmed) that the *Dinophysis* chloroplast is a kleptoplast—if so, *Dinophysis* chloroplasts wear out and *Dinophysis* species must continually engulf cryptophytes to obtain new chloroplasts to replace the old ones.

Chloroplast DNA

Chloroplasts have their own DNA, often abbreviated as ctDNA, or cpDNA. It is also known as the plastome. Its existence was first proved in 1962, and first sequenced in 1986—when two Japanese research teams sequenced the chloroplast DNA of liverwort and tobacco. Since then, hundreds of chloroplast DNAs from various species have been sequenced, but they are mostly those of land plants and green algae—glaucophytes, red algae, and other algal groups are extremely underrepresented, potentially introducing some bias in views of "typical" chloroplast DNA structure and content.

Molecular Structure

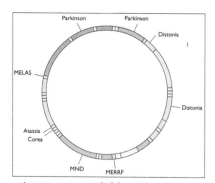

Chloroplast DNA Interactive gene map of chloroplast DNA from *Nicotiana tabacum*. Segments with labels on the inside reside on the B strand of DNA, segments with labels on the outside are on the A strand. Notches indicate introns.

With few exceptions, most chloroplasts have their entire chloroplast genome combined into a single large circular DNA molecule, typically 120,000–170,000 base pairs long. They can have a contour length of around 30–60 micrometers, and have a mass of about 80–130 million daltons.

While usually thought of as a circular molecule, there is some evidence that chloroplast DNA molecules more often take on a linear shape.

Inverted Repeats

Many chloroplast DNAs contain two *inverted repeats*, which separate a long single copy section (LSC) from a short single copy section (SSC). While a given pair of inverted repeats are rarely completely identical, they are always very similar to each other, apparently resulting from concerted evolution.

The inverted repeats vary wildly in length, ranging from 4,000 to 25,000 base pairs long each and containing as few as four or as many as over 150 genes. Inverted repeats in plants tend to be at the upper end of this range, each being 20,000–25,000 base pairs long.

The inverted repeat regions are highly conserved among land plants, and accumulate few mutations. Similar inverted repeats exist in the genomes of cyanobacteria and the other two chloroplast lineages (glaucophyta and rhodophyceae), suggesting that they predate the chloroplast, though some chloroplast DNAs have since lost or flipped the inverted repeats (making them direct repeats). It is possible that the inverted repeats help stabilize the rest of the chloroplast genome, as chloroplast DNAs which have lost some of the inverted repeat segments tend to get rearranged more.

Nucleoids

New chloroplasts may contain up to 100 copies of their DNA, though the number of chloroplast DNA copies decreases to about 15–20 as the chloroplasts age. They are usually packed into nucleoids, which can contain several identical chloroplast DNA rings. Many nucleoids can be found in each chloroplast. In primitive red algae, the chloroplast DNA nucleoids are clustered in the center of the chloroplast, while in green plants and green algae, the nucleoids are dispersed throughout the stroma.

Though chloroplast DNA is not associated with true histones, in red algae, similar proteins that tightly pack each chloroplast DNA ring into a nucleoid have been found.

DNA Repair

In chloroplasts of the moss *Physcomitrella patens*, the DNA mismatch repair protein

Msh1 interacts with the recombinational repair proteins RecA and RecG to maintain chloroplast genome stability. In chloroplasts of the plant *Arabidopsis thaliana* the RecA protein maintains the integrity of the chloroplast's DNA by a process that likely involves the recombinational repair of DNA damage.

DNA Replication

Leading Model of cpDNA Replication

The mechanism for chloroplast DNA (cpDNA) replication has not been conclusively determined, but two main models have been proposed. Scientists have attempted to observe chloroplast replication via electron microscopy since the 1970s. The results of the microscopy experiments led to the idea that chloroplast DNA replicates using a double displacement loop (D-loop). As the D-loop moves through the circular DNA, it adopts a theta intermediary form, also known as a Cairns replication intermediate, and completes replication with a rolling circle mechanism. Transcription starts at specific points of origin. Multiple replication forks open up, allowing replication machinery to transcribe the DNA. As replication continues, the forks grow and eventually converge. The new cpDNA structures separate, creating daughter cpDNA chromosomes.

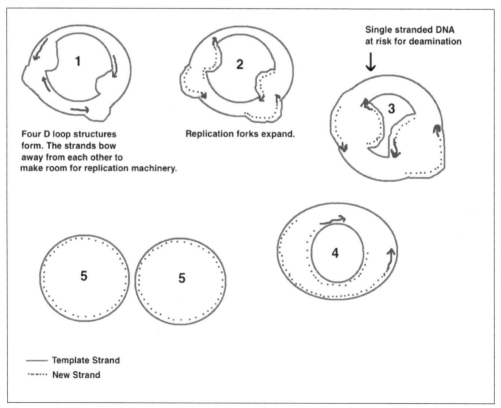

Chloroplast DNA replication via multiple D loop mechanisms. Adapted from Krishnan NM, Rao BJ's paper "A comparative approach to elucidate chloroplast genome replication."

In addition to the early microscopy experiments, this model is also supported by the amounts of deamination seen in cpDNA. Deamination occurs when an amino group is lost and is a mutation that often results in base changes. When adenine is deaminated, it becomes hypoxanthine. Hypoxanthine can bind to cytosine, and when the XC base pair is replicated, it becomes a GC (thus, an A → G base change).

Original DNA Strand

...CCATGCATGGATC...

Deamination of an Adenine

...CCATGCATGGATC...
↓
...CCHTGCATGGATC...

During Replication, H pairs with C

...CCHTGCATGGATC...
...GGCACGTACCTAG...

When Replicated Again, C pairs with G

...GGCACGTACCTAG...
...CCGTGCATGGATC...

Over time, base changes in the DNA sequence can arise from deamination mutations. When adenine is deaminated, it becomes hypoxanthine, which can pair with cytosine. During replication, the cytosine will pair with guanine, causing an A --> G base change.

Deamination

In cpDNA, there are several A → G deamination gradients. DNA becomes susceptible to deamination events when it is single stranded. When replication forks form, the strand not being copied is single stranded, and thus at risk for A → G deamination. Therefore, gradients in deamination indicate that replication forks were most likely present and the direction that they initially opened (the highest gradient is most likely nearest the start site because it was single stranded for the longest amount of time). This mechanism is still the leading theory today; however, a second theory suggests that most cpDNA is actually linear and replicates through homologous recombination. It further contends that only a minority of the genetic material is kept in circular chromosomes while the rest is in branched, linear, or other complex structures.

Alternative Model of Replication

One of competing model for cpDNA replication asserts that most cpDNA is linear and participates in homologous recombination and replication structures similar to the linear and circular DNA structures of bacteriophage T4. It has been established that some plants have linear cpDNA, such as maize, and that more species still contain complex structures that scientists do not yet understand. When the original experiments on cpDNA were performed, scientists did notice linear structures; however, they attributed these linear forms to broken circles. If the branched and complex structures seen in cpDNA experiments are real and not artifacts of concatenated circular DNA or broken circles, then a D-loop mechanism of replication is insufficient to explain how those structures would replicate. At the same time, homologous recombination does not expand the multiple A G gradients seen in plastomes. Because of the failure to explain the deamination gradient as well as the numerous plant species that have been shown to have circular cpDNA, the predominant theory continues to hold that most cpDNA is circular and most likely replicates via a D loop mechanism.

Gene Content and Protein Synthesis

The chloroplast genome most commonly includes around 100 genes that code for a variety of things, mostly to do with the protein pipeline and photosynthesis. As in prokaryotes, genes in chloroplast DNA are organized into operons. Unlike prokaryotic DNA molecules, chloroplast DNA molecules contain introns (plant mitochondrial DNAs do too, but not human mtDNAs). Among land plants, the contents of the chloroplast genome are fairly similar.

Chloroplast Genome Reduction and Gene Transfer

Over time, many parts of the chloroplast genome were transferred to the nuclear genome of the host, a process called *endosymbiotic gene transfer*. As a result, the chloroplast genome is heavily reduced compared to that of free-living cyanobacteria. Chloroplasts may contain 60–100 genes whereas cyanobacteria often have more than 1500 genes in their genome. Recently, a plastid without a genome was found, demonstrating chloroplasts can lose their genome during endosymbiotic the gene transfer process.

Endosymbiotic gene transfer is how we know about the lost chloroplasts in many CASH lineages. Even if a chloroplast is eventually lost, the genes it donated to the former host's nucleus persist, providing evidence for the lost chloroplast's existence. For example, while diatoms (a heterokontophyte) now have a red algal derived chloroplast, the presence of many green algal genes in the diatom nucleus provide evidence that the diatom ancestor had a green algal derived chloroplast at some point, which was subsequently replaced by the red chloroplast.

In land plants, some 11–14% of the DNA in their nuclei can be traced back to the chloroplast, up to 18% in *Arabidopsis*, corresponding to about 4,500 protein-coding genes. There have been a few recent transfers of genes from the chloroplast DNA to the nuclear genome in land plants.

Of the approximately 3000 proteins found in chloroplasts, some 95% of them are encoded by nuclear genes. Many of the chloroplast's protein complexes consist of subunits from both the chloroplast genome and the host's nuclear genome. As a result, protein synthesis must be coordinated between the chloroplast and the nucleus. The chloroplast is mostly under nuclear control, though chloroplasts can also give out signals regulating gene expression in the nucleus, called *retrograde signaling*.

Protein Synthesis

Protein synthesis within chloroplasts relies on two RNA polymerases. One is coded by the chloroplast DNA, the other is of nuclear origin. The two RNA polymerases may recognize and bind to different kinds of promoters within the chloroplast genome. The ribosomes in chloroplasts are similar to bacterial ribosomes.

Protein Targeting and Import

Because so many chloroplast genes have been moved to the nucleus, many proteins that would originally have been translated in the chloroplast are now synthesized in the cytoplasm of the plant cell. These proteins must be directed back to the chloroplast, and imported through at least two chloroplast membranes.

Curiously, around half of the protein products of transferred genes aren't even targeted back to the chloroplast. Many became exaptations, taking on new functions like participating in cell division, protein routing, and even disease resistance. A few chloroplast genes found new homes in the mitochondrial genome—most became nonfunctional pseudogenes, though a few tRNA genes still work in the mitochondrion. Some transferred chloroplast DNA protein products get directed to the secretory pathway though many secondary plastids are bounded by an outermost membrane derived from the host's cell membrane, and therefore topologically outside of the cell, because to reach the chloroplast from the cytosol, you have to cross the cell membrane, just like if you were headed for the extracellular space. In those cases, chloroplast-targeted proteins do initially travel along the secretory pathway.

Because the cell acquiring a chloroplast already had mitochondria (and peroxisomes, and a cell membrane for secretion), the new chloroplast host had to develop a unique protein targeting system to avoid having chloroplast proteins being sent to the wrong organelle.

The two ends of a polypeptide are called the N-terminus, or *amino end*, and the C-terminus, or *carboxyl end*. This polypeptide has four amino acids linked together. At the left is the N-terminus, with its amino (H_2N) group in green. The blue C-terminus, with its carboxyl group (CO_2H) is at the right.

In most, but not all cases, nuclear-encoded chloroplast proteins are translated with a *cleavable transit peptide* that's added to the N-terminus of the protein precursor. Sometimes the transit sequence is found on the C-terminus of the protein, or within the functional part of the protein.

Transport Proteins and Membrane Translocons

After a chloroplast polypeptide is synthesized on a ribosome in the cytosol, an enzyme specific to chloroplast proteins phosphorylates, or adds a phosphate group to many (but not all) of them in their transit sequences. Phosphorylation helps many proteins bind the polypeptide, keeping it from folding prematurely. This is important because it prevents chloroplast proteins from assuming their active form and carrying out their chloroplast functions in the wrong place—the cytosol. At the same time, they have to keep just enough shape so that they can be recognized by the chloroplast. These proteins also help the polypeptide get imported into the chloroplast.

From here, chloroplast proteins bound for the stroma must pass through two protein complexes—the TOC complex, or translocon on the outer chloroplast membrane, and the TIC translocon, or translocon on the inner chloroplast membrane translocon. Chloroplast polypeptide chains probably often travel through the two complexes at the same time, but the TIC complex can also retrieve preproteins lost in the intermembrane space.

Structure

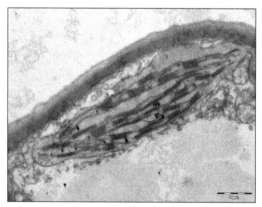

Transmission electron microscope image of a chloroplast. Grana of thylakoids and their connecting lamellae are clearly visible.

In land plants, chloroplasts are generally lens-shaped, 3–10 µm in diameter and 1–3 µm thick. Corn seedling chloroplasts are ≈20 µm³ in volume. Greater diversity in chloroplast shapes exists among the algae, which often contain a single chloroplast that can be shaped like a net (e.g., *Oedogonium*), a cup (e.g., *Chlamydomonas*), a ribbon-like spiral around the edges of the cell (e.g., *Spirogyra*), or slightly twisted bands at the cell edges (e.g., *Sirogonium*). Some algae have two chloroplasts in each cell; they are star-shaped in *Zygnema*, or may follow the shape of half the cell in order Desmidiales. In some algae, the chloroplast takes up most of the cell, with pockets for the nucleus and other organelles, for example, some species of *Chlorella* have a cup-shaped chloroplast that occupies much of the cell.

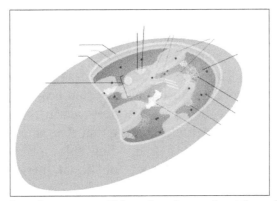

Chloroplast ultrastructure Chloroplasts have at least three distinct membrane systems, and a variety of things can be found in their stroma.

All chloroplasts have at least three membrane systems—the outer chloroplast membrane, the inner chloroplast membrane, and the thylakoid system. Chloroplasts that are the product of secondary endosymbiosis may have additional membranes surrounding these three. Inside the outer and inner chloroplast membranes is the chloroplast stroma, a semi-gel-like fluid that makes up much of a chloroplast's volume, and in which the thylakoid system floats.

There are some common misconceptions about the outer and inner chloroplast membranes. The fact that chloroplasts are surrounded by a double membrane is often cited as evidence that they are the descendants of endosymbiotic cyanobacteria. This is often interpreted as meaning the outer chloroplast membrane is the product of the host's cell membrane infolding to form a vesicle to surround the ancestral cyanobacterium—which is not true—both chloroplast membranes are homologous to the cyanobacterium's original double membranes.

The chloroplast double membrane is also often compared to the mitochondrial double membrane. This is not a valid comparison—the inner mitochondria membrane is used to run proton pumps and carry out oxidative phosphorylation across to generate ATP energy. The only chloroplast structure that can considered analogous to it is the internal thylakoid system. Even so, in terms of "in-out", the direction of chloroplast H+ ion flow is in the opposite direction compared to oxidative phosphorylation in mitochondria. In addition, in terms of function, the inner chloroplast membrane, which regulates metabolite passage and synthesizes some materials, has no counterpart in the mitochondrion.

Outer Chloroplast Membrane

The outer chloroplast membrane is a semi-porous membrane that small molecules and

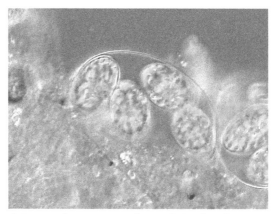

ions can easily diffuse across. However, it is not permeable to larger proteins, so chloroplast polypeptides being synthesized in the cell cytoplasm must be transported across the outer chloroplast membrane by the TOC complex, or *translocon on the outer chloroplast* membrane.

The chloroplast membranes sometimes protrude out into the cytoplasm, forming a stromule, or stroma-containing tubule. Stromules are very rare in chloroplasts, and are much more common in other plastids like chromoplasts and amyloplasts in petals and roots, respectively. They may exist to increase the chloroplast's surface area for cross-membrane transport, because they are often branched and tangled with the endoplasmic reticulum. When they were first observed in 1962, some plant biologists dismissed the structures as artifactual, claiming that stromules were just oddly shaped chloroplasts with constricted regions or dividing chloroplasts. However, there is a growing body of evidence that stromules are functional, integral features of plant cell plastids, not merely artifacts.

Intermembrane Space and Peptidoglycan Wall

Usually, a thin intermembrane space about 10–20 nanometers thick exists between the outer and inner chloroplast membranes.

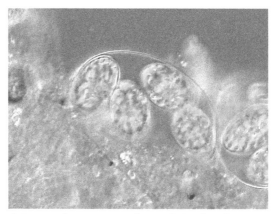

Instead of an intermembrane space, glaucophyte algae have a peptidoglycan wall between their inner and outer chloroplast membranes.

Glaucophyte algal chloroplasts have a peptidoglycan layer between the chloroplast membranes. It corresponds to the peptidoglycan cell wall of their cyanobacterial ancestors, which is located between their two cell membranes. These chloroplasts are called *muroplasts*. Other chloroplasts have lost the cyanobacterial wall, leaving an intermembrane space between the two chloroplast envelope membranes.

Inner Chloroplast Membrane

The inner chloroplast membrane borders the stroma and regulates passage of materials in and out of the chloroplast. After passing through the TOC complex in the outer chloroplast membrane, polypeptides must pass through the TIC complex *(translocon on the inner chloroplast membrane)* which is located in the inner chloroplast membrane.

In addition to regulating the passage of materials, the inner chloroplast membrane is where fatty acids, lipids, and carotenoids are synthesized.

Peripheral Reticulum

Some chloroplasts contain a structure called the chloroplast peripheral reticulum. It is often found in the chloroplasts of C4 plants, though it has also been found in some C_3 angiosperms, and even some gymnosperms. The chloroplast peripheral reticulum consists of a maze of membranous tubes and vesicles continuous with the inner chloroplast membrane that extends into the internal stromal fluid of the chloroplast. Its purpose is thought to be to increase the chloroplast's surface area for cross-membrane transport between its stroma and the cell cytoplasm. The small vesicles sometimes observed may serve as transport vesicles to shuttle stuff between the thylakoids and intermembrane space.

Stroma

The protein-rich, alkaline, aqueous fluid within the inner chloroplast membrane and outside of the thylakoid space is called the stroma, which corresponds to the cytosol of the original cyanobacterium. Nucleoids of chloroplast DNA, chloroplast ribosomes, the thylakoid system with plastoglobuli, starch granules, and many proteins can be found floating around in it. The Calvin cycle, which fixes CO2 into G3P takes place in the stroma.

Chloroplast Ribosomes

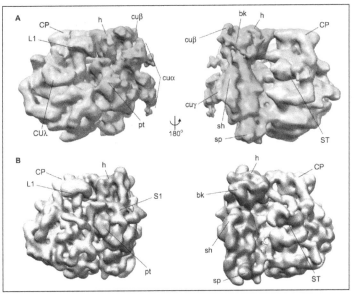

Chloroplast ribosomes Comparison of a chloroplast ribosome (green) and a bacterial ribosome (yellow). Important features common to both ribosomes and chloroplast-unique features are labeled.

Chloroplasts have their own ribosomes, which they use to synthesize a small fraction of their proteins. Chloroplast ribosomes are about two-thirds the size of cytoplasmic ribosomes (around 17 nm vs 25 nm). They take mRNAs transcribed from the chloroplast DNA and translate them into protein. While similar to bacterial ribosomes, chloroplast translation is more complex than in bacteria, so chloroplast ribosomes include some chloroplast-unique features. Small subunit ribosomal RNAs in several Chlorophyta and euglenid chloroplasts lack motifs for shine-dalgarno sequence recognition, which is considered essential for translation initiation in most chloroplasts and prokaryotes. Such loss is also rarely observed in other plastids and prokaryotes.

Plastoglobuli

Plastoglobuli (singular *plastoglobulus*, sometimes spelled *plastoglobule(s)*), are spherical bubbles of lipids and proteins about 45–60 nanometers across. They are surrounded by a lipid monolayer. Plastoglobuli are found in all chloroplasts, but become more common when the chloroplast is under oxidative stress, or when it ages and transitions into a gerontoplast. Plastoglobuli also exhibit a greater size variation under these conditions. They are also common in etioplasts, but decrease in number as the etioplasts mature into chloroplasts.

Plastoglubuli contain both structural proteins and enzymes involved in lipid synthesis and metabolism. They contain many types of lipids including plastoquinone, vitamin E, carotenoids and chlorophylls.

Plastoglobuli were once thought to be free-floating in the stroma, but it is now thought that they are permanently attached either to a thylakoid or to another plastoglobulus attached to a thylakoid, a configuration that allows a plastoglobulus to exchange its contents with the thylakoid network. In normal green chloroplasts, the vast majority of plastoglobuli occur singularly, attached directly to their parent thylakoid. In old or stressed chloroplasts, plastoglobuli tend to occur in linked groups or chains, still always anchored to a thylakoid.

Plastoglobuli form when a bubble appears between the layers of the lipid bilayer of the thylakoid membrane, or bud from existing plastoglubuli—though they never detach and float off into the stroma. Practically all plastoglobuli form on or near the highly curved edges of the thylakoid disks or sheets. They are also more common on stromal thylakoids than on granal ones.

Starch Granules

Starch granules are very common in chloroplasts, typically taking up 15% of the organelle's volume, though in some other plastids like amyloplasts, they can be big enough to distort the shape of the organelle. Starch granules are simply accumulations of starch in the stroma, and are not bounded by a membrane.

Starch granules appear and grow throughout the day, as the chloroplast synthesizes sugars, and are consumed at night to fuel respiration and continue sugar export into the phloem, though in mature chloroplasts, it is rare for a starch granule to be completely consumed or for a new granule to accumulate.

Starch granules vary in composition and location across different chloroplast lineages. In red algae, starch granules are found in the cytoplasm rather than in the chloroplast. In C4 plants, mesophyll chloroplasts, which do not synthesize sugars, lack starch granules.

RuBisCO

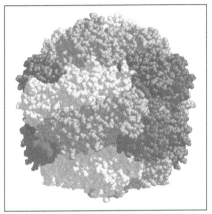

RuBisCO, shown here in a space-filling model, is the main enzyme responsible for carbon fixation in chloroplasts.

The chloroplast stroma contains many proteins, though the most common and important is RuBisCO, which is probably also the most abundant protein on the planet. RuBisCO is the enzyme that fixes CO_2 into sugar molecules. In C_3 plants, RuBisCO is abundant in all chloroplasts, though in C_4 plants, it is confined to the bundle sheath chloroplasts, where the Calvin cycle is carried out in C_4 plants.

Pyrenoids

The chloroplasts of some hornworts and algae contain structures called pyrenoids. They are not found in higher plants. Pyrenoids are roughly spherical and highly refractive bodies which are a site of starch accumulation in plants that contain them. They consist of a matrix opaque to electrons, surrounded by two hemispherical starch plates. The starch is accumulated as the pyrenoids mature. In algae with carbon concentrating mechanisms, the enzyme RuBisCO is found in the pyrenoids. Starch can also accumulate around the pyrenoids when CO_2 is scarce. Pyrenoids can divide to form new pyrenoids, or be produced "de novo".

Thylakoid System

Suspended within the chloroplast stroma is the thylakoid system, a highly dynamic collection of membranous sacks called thylakoids where chlorophyll is found and the light reactions of photosynthesis happen. In most vascular plant chloroplasts, the thylakoids are arranged in stacks called grana, though in certain C_4 plant chloroplasts and some algal chloroplasts, the thylakoids are free floating.

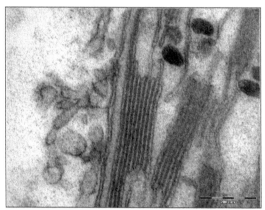

Transmission electron microscope image of some thylakoids arranged in grana stacks and lamellæ. Plastoglobuli (dark blobs) are also present.

Granal Structure

Using a light microscope, it is just barely possible to see tiny green granules—which were named grana. With electron microscopy, it became possible to see the thylakoid system in more detail, revealing it to consist of stacks of flat thylakoids which made up the grana, and long interconnecting stromal thylakoids which linked different grana. In the transmission electron microscope, thylakoid membranes appear as alternating light-and-dark bands, 8.5 nanometers thick.

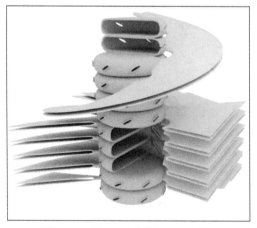

Granum structure: The prevailing model for granal structure is a stack of granal thylakoids linked by helical stromal thylakoids that wrap around the grana stacks and form large sheets that connect different grana.

For a long time, the three-dimensional structure of the thylakoid system has been unknown or disputed. One model has the granum as a stack of thylakoids linked by helical stromal thylakoids; the other has the granum as a single folded thylakoid connected in a "hub and spoke" way to other grana by stromal thylakoids. While the thylakoid system is still commonly depicted according to the folded thylakoid model, it was determined in 2011 that the stacked and helical thylakoids model is correct.

In the helical thylakoid model, grana consist of a stack of flattened circular granal thylakoids that resemble pancakes. Each granum can contain anywhere from two to a hundred thylakoids, though grana with 10–20 thylakoids are most common. Wrapped around the grana are helicoid stromal thylakoids, also known as frets or lamellar thylakoids. The helices ascend at an angle of 20–25°, connecting to each granal thylakoid at a bridge-like slit junction. The helicoids may extend as large sheets that link multiple grana, or narrow to tube-like bridges between grana. While different parts of the thylakoid system contain different membrane proteins, the thylakoid membranes are continuous and the thylakoid space they enclose form a single continuous labyrinth.

Thylakoids

Thylakoids (sometimes spelled *thylakoïds*), are small interconnected sacks which contain the membranes that the light reactions of photosynthesis take place on.

Embedded in the thylakoid membranes are important protein complexes which carry out the light reactions of photosynthesis. Photosystem II and photosystem I contain light-harvesting complexes with chlorophyll and carotenoids that absorb light energy and use it to energize electrons. Molecules in the thylakoid membrane use the energized electrons to pump hydrogen ions into the thylakoid space, decreasing the pH and turning it acidic. ATP synthase is a large protein complex that harnesses the concentration gradient of the hydrogen ions in the thylakoid space to generate ATP energy as the hydrogen ions flow back out into the stroma—much like a dam turbine.

There are two types of thylakoids—granal thylakoids, which are arranged in grana, and stromal thylakoids, which are in contact with the stroma. Granal thylakoids are pancake-shaped circular disks about 300–600 nanometers in diameter. Stromal thylakoids are helicoid sheets that spiral around grana. The flat tops and bottoms of granal thylakoids contain only the relatively flat photosystem II protein complex. This allows them to stack tightly, forming grana with many layers of tightly appressed membrane, called granal membrane, increasing stability and surface area for light capture.

In contrast, photosystem I and ATP synthase are large protein complexes which jut out into the stroma. They can't fit in the appressed granal membranes, and so are found in the stromal thylakoid membrane—the edges of the granal thylakoid disks and the stromal thylakoids. These large protein complexes may act as spacers between the sheets of stromal thylakoids.

The number of thylakoids and the total thylakoid area of a chloroplast is influenced by light exposure. Shaded chloroplasts contain larger and more grana with more thylakoid membrane area than chloroplasts exposed to bright light, which have smaller and fewer grana and less thylakoid area. Thylakoid extent can change within minutes of light exposure or removal.

Pigments and Chloroplast Colors

Inside the photosystems embedded in chloroplast thylakoid membranes are various photosynthetic pigments, which absorb and transfer light energy. The types of pigments found are different in various groups of chloroplasts, and are responsible for a wide variety of chloroplast colorations.

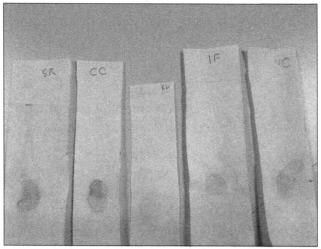

Paper chroma-tography of some spinach leaf extract shows the various pigments present in their chloroplasts.

Chlorophylls

Chlorophyll *a* is found in all chloroplasts, as well as their cyanobacterial ancestors. Chlorophyll *a* is a blue-green pigment partially responsible for giving most cyanobacteria and chloroplasts their color. Other forms of chlorophyll exist, such as the accessory pigments chlorophyll *b*, chlorophyll *c*, chlorophyll *d*, and chlorophyll *f*.

Chlorophyll *b* is an olive green pigment found only in the chloroplasts of plants, green algae, any secondary chloroplasts obtained through the secondary endosymbiosis of a green alga, and a few cyanobacteria. It is the chlorophylls *a* and *b* together that make most plant and green algal chloroplasts green.

Chlorophyll *c* is mainly found in secondary endosymbiotic chloroplasts that originated from a red alga, although it is not found in chloroplasts of red algae themselves. Chlorophyll *c* is also found in some green algae and cyanobacteria. Chlorophylls *d* and *f* are pigments found only in some cyanobacteria.

Carotenoids

In addition to chlorophylls, another group of yellow–orange pigments called carotenoids are also found in the photosystems. There are about thirty photosynthetic carotenoids. They help transfer and dissipate excess energy, and their bright colors sometimes override the chlorophyll green, like during the fall, when the leaves of some land plants change color. β-carotene is a bright red-orange carotenoid found in nearly all chloroplasts, like chlorophyll a. Xanthophylls, especially the orange-red zeaxanthin, are also common. Many other forms of carotenoids exist that are only found in certain groups of chloroplasts.

Delesseria sanguinea, a red alga, has chloroplasts that contain red pigments like phycoerytherin that mask their blue-green chlorophyll a.

Phycobilins

Phycobilins are a third group of pigments found in cyanobacteria, and glaucophyte, red algal, and cryptophyte chloroplasts. Phycobilins come in all colors, though phycoerytherin is one of the pigments that makes many red algae red. Phycobilins often organize into relatively large protein complexes about 40 nanometers across called phycobilisomes. Like photosystem I and ATP synthase, phycobilisomes jut into the stroma, preventing thylakoid stacking in red algal chloroplasts. Cryptophyte chloroplasts and some cyanobacteria don't have their phycobilin pigments organized into phycobilisomes, and keep them in their thylakoid space instead.

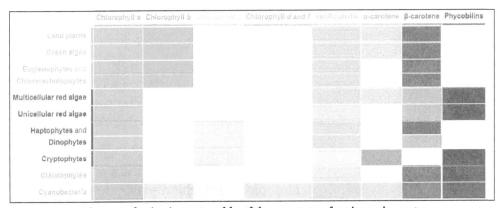

Photosynthetic pigments table of the presence of various pigments across chloroplast groups. Colored cells represent pigment presence.

Specialized chloroplasts in C$_4$ plants

To fix carbon dioxide into sugar molecules in the process of photosynthesis, chloroplasts use an enzyme called RuBisCO. RuBisCO has a problem—it has trouble distinguishing between carbon dioxide and oxygen, so at high oxygen concentrations, RuBisCO starts accidentally adding oxygen to sugar precursors. This has the end result of ATP energy being wasted and CO$_2$ being released, all with no sugar being produced. This is a big problem, since O$_2$ is produced by the initial light reactions of photosynthesis, causing issues down the line in the Calvin cycle which uses RuBisCO.

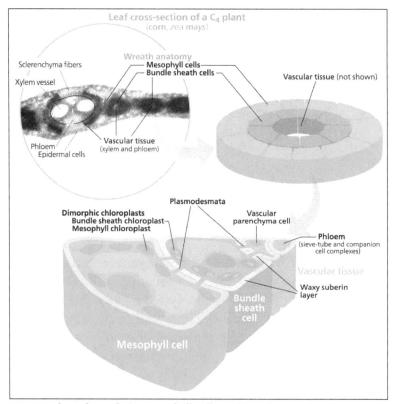

Many C$_4$ plants have their mesophyll cells and bundle sheath cells arranged radially around their leaf veins. The two types of cells contain different types of chloroplasts specialized for a particular part of photosynthesis.

C$_4$ plants evolved a way to solve this—by spatially separating the light reactions and the Calvin cycle. The light reactions, which store light energy in ATP and NADPH, are done in the mesophyll cells of a C$_4$ leaf. The Calvin cycle, which uses the stored energy to make sugar using RuBisCO, is done in the bundle sheath cells, a layer of cells surrounding a vein in a leaf.

As a result, chloroplasts in C$_4$ mesophyll cells and bundle sheath cells are specialized for each stage of photosynthesis. In mesophyll cells, chloroplasts are specialized for the light reactions, so they lack RuBisCO, and have normal grana and thylakoids, which they use to make ATP and NADPH, as well as oxygen. They store CO$_2$ in a four-carbon compound, which is why the process is called C$_4$ *photosynthesis*. The four-carbon compound is then transported to the bundle sheath chloroplasts, where it drops off CO$_2$ and returns to the mesophyll. Bundle sheath chloroplasts do not carry out the light reactions, preventing oxygen from building up in them and disrupting RuBisCO activity. Because of this, they lack thylakoids organized into grana stacks—though bundle sheath chloroplasts still have free-floating thylakoids in the stroma where they still carry out cyclic electron flow, a light-driven method of synthesizing ATP to power the Calvin cycle without generating oxygen. They lack photosystem II, and only have photosystem I—the only protein complex needed for cyclic electron flow. Because the job of bundle sheath chloroplasts is to carry out the Calvin cycle and make sugar, they often contain large starch grains.

Both types of chloroplast contain large amounts of chloroplast peripheral reticulum, which they use to get more surface area to transport stuff in and out of them. Mesophyll chloroplasts have a little more peripheral reticulum than bundle sheath chloroplasts.

Distribution in a Plant

Not all cells in a multicellular plant contain chloroplasts. All green parts of a plant contain chloroplasts—the chloroplasts, or more specifically, the chlorophyll in them are what make the photosynthetic parts of a plant green. The plant cells which contain chloroplasts are usually parenchyma cells, though chloroplasts can also be found in collenchyma tissue. A plant cell which contains chloroplasts is known as a chlorenchyma cell. A typical chlorenchyma cell of a land plant contains about 10 to 100 chloroplasts.

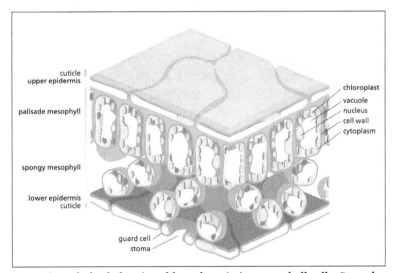

A cross section of a leaf, showing chloroplasts in its mesophyll cells. Stomal guard cells also have chloroplasts, though much fewer than mesophyll cells.

In some plants such as cacti, chloroplasts are found in the stems, though in most plants, chloroplasts are concentrated in the leaves. One square millimeter of leaf tissue can contain half a million chloroplasts. Within a leaf, chloroplasts are mainly found in the mesophyll layers of a leaf, and the guard cells of stomata. Palisade mesophyll cells can contain 30–70 chloroplasts per cell, while stomatal guard cells contain only around 8–15 per cell, as well as much less chlorophyll. Chloroplasts can also be found in the bundle sheath cells of a leaf, especially in C_4 plants, which carry out the Calvin cycle in their bundle sheath cells. They are often absent from the epidermis of a leaf.

Cellular Location

Chloroplast Movement

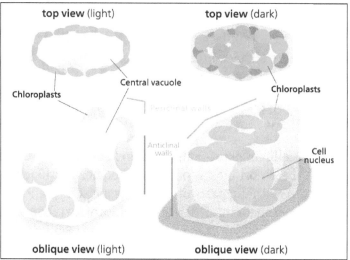

When chloroplasts are exposed to direct sunlight, they stack along the anticlinal cell walls to minimize exposure. In the dark they spread out in sheets along the periclinal walls to maximize light absorption.

The chloroplasts of plant and algal cells can orient themselves to best suit the available light. In low-light conditions, they will spread out in a sheet—maximizing the surface area to absorb light. Under intense light, they will seek shelter by aligning in vertical columns along the plant cell's cell wall or turning sideways so that light strikes them edge-on. This reduces exposure and protects them from photooxidative damage. This ability to distribute chloroplasts so that they can take shelter behind each other or spread out may be the reason why land plants evolved to have many small chloroplasts instead of a few big ones. Chloroplast movement is considered one of the most closely regulated stimulus-response systems that can be found in plants. Mitochondria have also been observed to follow chloroplasts as they move.

In higher plants, chloroplast movement is run by phototropins, blue light photoreceptors also responsible for plant phototropism. In some algae, mosses, ferns, and flowering plants, chloroplast movement is influenced by red light in addition to blue light, though very long red wavelengths inhibit movement rather than speeding it up. Blue light generally causes chloroplasts to seek shelter, while red light draws them out to maximize light absorption.

Studies of *Vallisneria gigantea*, an aquatic flowering plant, have shown that chloroplasts can get moving within five minutes of light exposure, though they don't initially show any net directionality. They may move along microfilament tracks, and the fact that the microfilament mesh changes shape to form a honeycomb structure surrounding the chloroplasts after they have moved suggests that microfilaments may help to anchor chloroplasts in place.

Function and chemistry

Guard Cell Chloroplasts

Unlike most epidermal cells, the guard cells of plant stomata contain relatively well-developed chloroplasts. However, exactly what they do is controversial.

Plant Innate Immunity

Plants lack specialized immune cells—all plant cells participate in the plant immune response. Chloroplasts, along with the nucleus, cell membrane, and endoplasmic reticulum, are key players in pathogen defense. Due to its role in a plant cell's immune response, pathogens frequently target the chloroplast.

Plants have two main immune responses—the hypersensitive response, in which infected cells seal themselves off and undergo programmed cell death, and systemic acquired resistance, where infected cells release signals warning the rest of the plant of a pathogen's presence. Chloroplasts stimulate both responses by purposely damaging their photosynthetic system, producing reactive oxygen species. High levels of reactive oxygen species will cause the hypersensitive response. The reactive oxygen species also directly kill any pathogens within the cell. Lower levels of reactive oxygen species initiate systemic acquired resistance, triggering defense-molecule production in the rest of the plant.

In some plants, chloroplasts are known to move closer to the infection site and the nucleus during an infection.

Chloroplasts can serve as cellular sensors. After detecting stress in a cell, which might be due to a pathogen, chloroplasts begin producing molecules like salicylic acid, jasmonic acid, nitric oxide and reactive oxygen species which can serve as defense-signals. As cellular signals, reactive oxygen species are unstable molecules, so they probably don't leave the chloroplast, but instead pass on their signal to an unknown second messenger molecule. All these molecules initiate retrograde signaling—signals from the chloroplast that regulate gene expression in the nucleus.

In addition to defense signaling, chloroplasts, with the help of the peroxisomes, help synthesize an important defense molecule, jasmonate. Chloroplasts synthesize all the fatty acids in a plant cell—linoleic acid, a fatty acid, is a precursor to jasmonate.

Photosynthesis

One of the main functions of the chloroplast is its role in photosynthesis, the process by which light is transformed into chemical energy, to subsequently produce food in the form of sugars. Water (H_2O) and carbon dioxide (CO_2) are used in photosynthesis, and sugar and oxygen (O_2) is made, using light energy. Photosynthesis is divided into two stages—the light reactions, where water is split to produce oxygen, and the dark reactions, or Calvin cycle, which builds sugar molecules from carbon dioxide. The two phases are linked by the energy carriers adenosine triphosphate (ATP) and nicotinamide adenine dinucleotide phosphate ($NADP^+$).

Light Reactions

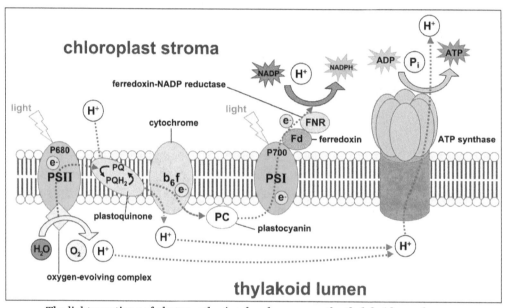

The light reactions of photosynthesis take place across the thylakoid membranes.

The light reactions take place on the thylakoid membranes. They take light energy and store it in NADPH, a form of $NADP^+$, and ATP to fuel the dark reactions.

Energy Carriers

ATP is the phosphorylated version of adenosine diphosphate (ADP), which stores energy in a cell and powers most cellular activities. ATP is the energized form, while ADP is the (partially) depleted form. $NADP^+$ is an electron carrier which ferries high energy electrons. In the light reactions, it gets reduced, meaning it picks up electrons, becoming NADPH.

Photophosphorylation

Like mitochondria, chloroplasts use the potential energy stored in an H+, or hydrogen ion gradient to generate ATP energy. The two photosystems capture light energy to energize electrons taken from water, and release them down an electron transport chain. The molecules between the photosystems harness the electrons' energy to pump hydrogen ions into the thylakoid space, creating a concentration gradient, with more hydrogen ions (up to a thousand times as many) inside the thylakoid system than in the stroma. The hydrogen ions in the thylakoid space then diffuse back down their concentration gradient, flowing back out into the stroma through ATP synthase. ATP synthase uses the energy from the flowing hydrogen ions to phosphorylate adenosine diphosphate into adenosine triphosphate, or ATP. Because chloroplast ATP synthase projects out into the stroma, the ATP is synthesized there, in position to be used in the dark reactions.

NADP⁺ Reduction

Electrons are often removed from the electron transport chains to charge $NADP^+$ with electrons, reducing it to NADPH. Like ATP synthase, ferredoxin-$NADP^+$ reductase, the enzyme that reduces $NADP^+$, releases the NADPH it makes into the stroma, right where it is needed for the dark reactions.

Because $NADP^+$ reduction removes electrons from the electron transport chains, they must be replaced—the job of photosystem II, which splits water molecules (H_2O) to obtain the electrons from its hydrogen atoms.

Cyclic Photophosphorylation

While photosystem II photolyzes water to obtain and energize new electrons, photosystem I simply reenergizes depleted electrons at the end of an electron transport chain. Normally, the reenergized electrons are taken by $NADP^+$, though sometimes they can flow back down more H^+-pumping electron transport chains to transport more hydrogen ions into the thylakoid space to generate more ATP. This is termed cyclic photophosphorylation because the electrons are recycled. Cyclic photophosphorylation is common in C4 plants, which need more ATP than NADPH.

Dark Reactions

The Calvin cycle, also known as the dark reactions, is a series of biochemical reactions that fixes CO_2 into G3P sugar molecules and uses the energy and electrons from the ATP and NADPH made in the light reactions. The Calvin cycle takes place in the stroma of the chloroplast.

While named *"the dark reactions"*, in most plants, they take place in the light, since the dark reactions are dependent on the products of the light reactions.

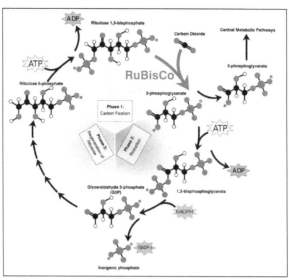

The Calvin cycle incorporates carbon dioxide into sugar molecules.

Carbon Fixation and G3P Synthesis

The Calvin cycle starts by using the enzyme RuBisCO to fix CO_2 into five-carbon Ribu-lose bisphosphate (RuBP) molecules. The result is unstable six-carbon molecules that immediately break down into three-carbon molecules called 3-phosphoglyceric acid, or 3-PGA. The ATP and NADPH made in the light reactions is used to convert the 3-PGA into glyceraldehyde-3-phosphate, or G3P sugar molecules. Most of the G3P molecules are recycled back into RuBP using energy from more ATP, but one out of every six pro-duced leaves the cycle—the end product of the dark reactions.

Sugars and Starches

Glyceraldehyde-3-phosphate can double up to form larger sugar molecules like glucose and fructose. These molecules are processed, and from them, the still larger sucrose, a disaccharide commonly known as table sugar, is made, though this process takes place outside of the chloroplast, in the cytoplasm.

Sucrose is made up of a glucose monomer (left), and a fructose monomer (right).

Alternatively, glucose monomers in the chloroplast can be linked together to make starch, which accumulates into the starch grains found in the chloroplast. Under condi-tions such as high atmospheric CO_2 concentrations, these starch grains may grow very large, distorting the grana and thylakoids. The starch granules displace the thylakoids, but leave them intact. Waterlogged roots can also cause starch buildup in the chloroplasts, possibly due to less sucrose being exported out of the chloroplast (or more accurately, the plant cell). This depletes a plant's free phosphate supply, which indirectly stimulates chloroplast starch synthesis. While linked to low photosynthesis rates, the starch grains themselves may not necessarily interfere significantly with the efficiency of photosynthesis, and might simply be a side effect of another photosynthesis-depressing factor.

Photorespiration

Photorespiration can occur when the oxygen concentration is too high. RuBisCO cannot distinguish between oxygen and carbon dioxide very well, so it can accidentally add O_2 instead of CO_2 to RuBP. This process reduces the efficiency of photosynthesis—it consumes ATP and oxygen, releases CO_2, and produces no sugar. It can waste up to half the carbon fixed by the Calvin cycle. Several mechanisms have evolved in different lineages that raise the carbon dioxide concentration relative to oxygen within the chloroplast, increasing the efficiency of photosynthesis. These mechanisms are called carbon dioxide concentrating mechanisms, or CCMs. These include Crassulacean acid metabolism, C_4 carbon fixation, and pyrenoids. Chloroplasts in C_4 plants are notable as they exhibit a distinct chloroplast dimorphism.

pH

Because of the H^+ gradient across the thylakoid membrane, the interior of the thylakoid is acidic, with a pH around 4, while the stroma is slightly basic, with a pH of around 8. The optimal stroma pH for the Calvin cycle is 8.1, with the reaction nearly stopping when the pH falls below 7.3.

CO_2 in water can form carbonic acid, which can disturb the pH of isolated chloroplasts, interfering with photosynthesis, even though CO_2 is used in photosynthesis. However, chloroplasts in living plant cells are not affected by this as much.

Chloroplasts can pump K^+ and H^+ ions in and out of themselves using a poorly understood light-driven transport system.

In the presence of light, the pH of the thylakoid lumen can drop up to 1.5 pH units, while the pH of the stroma can rise by nearly one pH unit.

Amino Acid Synthesis

Chloroplasts alone make almost all of a plant cell's amino acids in their stroma except the sulfur-containing ones like cysteine and methionine. Cysteine is made in the chloroplast (the proplastid too) but it is also synthesized in the cytosol and mitochondria, probably because it has trouble crossing membranes to get to where it is needed. The chloroplast is known to make the precursors to methionine but it is unclear whether the organelle carries out the last leg of the pathway or if it happens in the cytosol.

Other Nitrogen Compounds

Chloroplasts make all of a cell's purines and pyrimidines—the nitrogenous bases found in DNA and RNA. They also convert nitrite (NO_2^-) into ammonia (NH_3) which supplies the plant with nitrogen to make its amino acids and nucleotides.

Other Chemical Products

The plastid is the site of diverse and complex lipid synthesis in plants. The carbon used to form the majority of the lipid is from acetyl-CoA, which is the decarboxylation product of pyruvate. Pyruvate may enter the plastid from the cytosol by passive diffusion through the membrane after production in glycolysis. Pyruvate is also made in the plastid from phosphoenolpyruvate, a metabolite made in the cytosol from pyruvate or PGA. Acetate in the cytosol is unavailable for lipid biosynthesis in the plastid. The typical length of fatty acids produced in the plastid are 16 or 18 carbons, with 0-3 cis double bonds.

The biosynthesis of fatty acids from acetyl-CoA primarily requires two enzymes. Acetyl-CoA carboxylase creates malonyl-CoA, used in both the first step and the extension steps of synthesis. Fatty acid synthase (FAS) is a large complex of enzymes and cofactors including acyl carrier protein (ACP) which holds the acyl chain as it is synthesized. The initiation of synthesis begins with the condensation of malonyl-ACP with acetyl-CoA to produce ketobutyryl-ACP. 2 reductions involving the use of NADPH and one dehydration creates butyryl-ACP. Extension of the fatty acid comes from repeated cycles of malonyl-ACP condensation, reduction, and dehydration.

Other lipids are derived from the methyl-erythritol phosphate (MEP) pathway and consist of gibberelins, sterols, abscisic acid, phytol, and innumerable secondary metabolites.

Differentiation, Replication and Inheritance

Chloroplasts are a special type of a plant cell organelle called a plastid, though the two terms are sometimes used interchangeably. There are many other types of plastids, which carry out various functions. All chloroplasts in a plant are descended from undifferentiated proplastids found in the zygote, or fertilized egg. Proplastids are commonly found in an adult plant's apical meristems. Chloroplasts do not normally develop from proplastids in root tip meristems—instead, the formation of starch-storing amyloplasts is more common.

In shoots, proplastids from shoot apical meristems can gradually develop into chloroplasts in photosynthetic leaf tissues as the leaf matures, if exposed to the required light. This process involves invaginations of the inner plastid membrane, forming sheets of membrane that project into the internal stroma. These membrane sheets then fold to form thylakoids and grana.

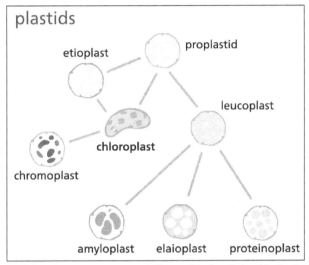

Plastid types Plants contain many different kinds of plastids in their cells.

If angiosperm shoots are not exposed to the required light for chloroplast formation, proplastids may develop into an etioplast stage before becoming chloroplasts. An etioplast is a plastid that lacks chlorophyll, and has inner membrane invaginations that form a lattice of tubes in their stroma, called a prolamellar body. While etioplasts lack chlorophyll, they have a yellow chlorophyll precursor stocked. Within a few minutes of light exposure, the prolamellar body begins to reorganize into stacks of thylakoids, and chlorophyll starts to be produced. This process, where the etioplast becomes a chloroplast, takes several hours. Gymnosperms do not require light to form chloroplasts.

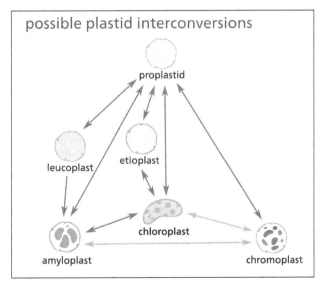

Many plastid interconversions are possible.

Light, however, does not guarantee that a proplastid will develop into a chloroplast. Whether a proplastid develops into a chloroplast some other kind of plastid is mostly controlled by the nucleus and is largely influenced by the kind of cell it resides in.

Plastid Interconversion

Plastid differentiation is not permanent, in fact many interconversions are possible. Chloroplasts may be converted to chromoplasts, which are pigment-filled plastids responsible for the bright colors seen in flowers and ripe fruit. Starch storing amyloplasts can also be converted to chromoplasts, and it is possible for proplastids to develop straight into chromoplasts. Chromoplasts and amyloplasts can also become chloroplasts, like what happens when a carrot or a potato is illuminated. If a plant is injured, or something else causes a plant cell to revert to a meristematic state, chloroplasts and other plastids can turn back into proplastids. Chloroplast, amyloplast, chromoplast, proplast, etc., are not absolute states—intermediate forms are common.

Chloroplast Division

Most chloroplasts in a photosynthetic cell do not develop directly from proplastids or etioplasts. In fact, a typical shoot meristematic plant cell contains only 7–20 proplastids. These proplastids differentiate into chloroplasts, which divide to create the 30–70 chloroplasts found in a mature photosynthetic plant cell. If the cell divides, chloroplast division provides the additional chloroplasts to partition between the two daughter cells.

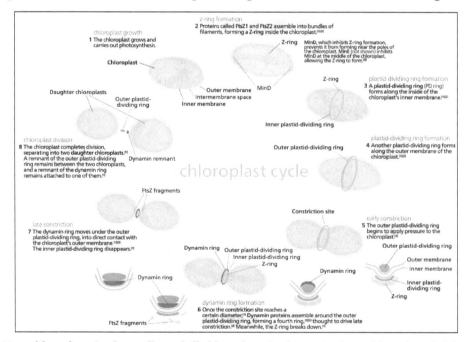

Most chloroplasts in plant cells, and all chloroplasts in algae arise from chloroplast division.

In single-celled algae, chloroplast division is the only way new chloroplasts are formed.

There is no proplastid differentiation—when an algal cell divides, its chloroplast divides along with it, and each daughter cell receives a mature chloroplast.

Almost all chloroplasts in a cell divide, rather than a small group of rapidly dividing chloroplasts. Chloroplasts have no definite S-phase—their DNA replication is not synchronized or limited to that of their host cells. Much of what we know about chloroplast division comes from studying organisms like *Arabidopsis* and the red alga *Cyanidioschyzon merolæ*.

The division process starts when the proteins FtsZ1 and FtsZ2 assemble into filaments, and with the help of a protein ARC6, form a structure called a Z-ring within the chloroplast's stroma. The Min system manages the placement of the Z-ring, ensuring that the chloroplast is cleaved more or less evenly. The protein MinD prevents FtsZ from linking up and forming filaments. Another protein ARC3 may also be involved, but it is not very well understood. These proteins are active at the poles of the chloroplast, preventing Z-ring formation there, but near the center of the chloroplast, MinE inhibits them, allowing the Z-ring to form.

Next, the two plastid-dividing rings, or PD rings form. The inner plastid-dividing ring is located in the inner side of the chloroplast's inner membrane, and is formed first. The outer plastid-dividing ring is found wrapped around the outer chloroplast membrane. It consists of filaments about 5 nanometers across, arranged in rows 6.4 nanometers apart, and shrinks to squeeze the chloroplast. This is when chloroplast constriction begins. In a few species like *Cyanidioschyzon merolæ*, chloroplasts have a third plastid-dividing ring located in the chloroplast's intermembrane space.

Late into the constriction phase, dynamin proteins assemble around the outer plastid-dividing ring, helping provide force to squeeze the chloroplast. Meanwhile, the Z-ring and the inner plastid-dividing ring break down. During this stage, the many chloroplast DNA plasmids floating around in the stroma are partitioned and distributed to the two forming daughter chloroplasts.

Later, the dynamins migrate under the outer plastid dividing ring, into direct contact with the chloroplast's outer membrane, to cleave the chloroplast in two daughter chloroplasts.

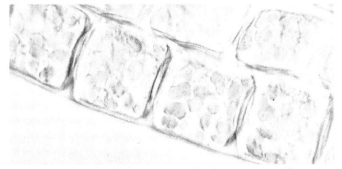

Chloroplast division: In this light micrograph of some moss chloroplasts, many dumbbell-shaped chloroplasts can be seen dividing. Grana are also just barely visible as small granules.

A remnant of the outer plastid dividing ring remains floating between the two daughter chloroplasts, and a remnant of the dynamin ring remains attached to one of the daughter chloroplasts.

Of the five or six rings involved in chloroplast division, only the outer plastid-dividing ring is present for the entire constriction and division phase—while the Z-ring forms first, constriction does not begin until the outer plastid-dividing ring forms.

Regulation

In species of algae that contain a single chloroplast, regulation of chloroplast division is extremely important to ensure that each daughter cell receives a chloroplast—chloroplasts can't be made from scratch. In organisms like plants, whose cells contain multiple chloroplasts, coordination is looser and less important. It is likely that chloroplast and cell division are somewhat synchronized, though the mechanisms for it are mostly unknown.

Light has been shown to be a requirement for chloroplast division. Chloroplasts can grow and progress through some of the constriction stages under poor quality green light, but are slow to complete division—they require exposure to bright white light to complete division. Spinach leaves grown under green light have been observed to contain many large dumbbell-shaped chloroplasts. Exposure to white light can stimulate these chloroplasts to divide and reduce the population of dumbbell-shaped chloroplasts.

Chloroplast Inheritance

Like mitochondria, chloroplasts are usually inherited from a single parent. Biparental chloroplast inheritance—where plastid genes are inherited from both parent plants—occurs in very low levels in some flowering plants.

Many mechanisms prevent biparental chloroplast DNA inheritance, including selective destruction of chloroplasts or their genes within the gamete or zygote, and chloroplasts from one parent being excluded from the embryo. Parental chloroplasts can be sorted so that only one type is present in each offspring.

Gymnosperms, such as pine trees, mostly pass on chloroplasts paternally, while flowering plants often inherit chloroplasts maternally. Flowering plants were once thought to only inherit chloroplasts maternally. However, there are now many documented cases of angiosperms inheriting chloroplasts paternally.

Angiosperms, which pass on chloroplasts maternally, have many ways to prevent paternal inheritance. Most of them produce sperm cells that do not contain any plastids. There are many other documented mechanisms that prevent paternal inheritance in these flowering plants, such as different rates of chloroplast replication within the embryo.

Among angiosperms, paternal chloroplast inheritance is observed more often in hybrids than in offspring from parents of the same species. This suggests that incompatible hybrid genes might interfere with the mechanisms that prevent paternal inheritance.

Transplastomic Plants

Recently, chloroplasts have caught attention by developers of genetically modified crops. Since, in most flowering plants, chloroplasts are not inherited from the male parent, transgenes in these plastids cannot be disseminated by pollen. This makes plastid transformation a valuable tool for the creation and cultivation of genetically modified plants that are biologically contained, thus posing significantly lower environmental risks. This biological containment strategy is therefore suitable for establishing the co-existence of conventional and organic agriculture. While the reliability of this mechanism has not yet been studied for all relevant crop species, recent results in tobacco plants are promising, showing a failed containment rate of transplastomic plants at 3 in 1,000,000.

CHLOROSOME

A chlorosome is a photosynthetic antenna complex found in green sulfur bacteria (GSB) and some green filamentous anoxygenic phototrophs (FAP) (Chloroflexaceae, Oscillochloridaceae; both members of Chlorflexia). They differ from other antenna complexes by their large size and lack of protein matrix supporting the photosynthetic pigments. Green sulfur bacteria are a group of organisms that generally live in extremely low-light environments, such as at depths of 100 metres in the Black Sea. The ability to capture light energy and rapidly deliver it to where it needs to go is essential to these bacteria, some of which see only a few photons of light per chlorophyll per day. To achieve this, the bacteria contain chlorosome structures, which contain up to 250,000 chlorophyll molecules. Chlorosomes are ellipsoidal bodies, in GSB their length varies from 100 to 200 nm, width of 50-100 nm and height of 15 - 30 nm, in FAP the chlorosomes are somewhat smaller.

Structure

Chlorosome shape can vary between species, with some species containing ellipsoidal shaped chlorosomes and others containing conical or irregular shaped chlorosomes. Inside green sulfur bacteria, the chlorosomes are attached to type-I reaction centers in the cell membrane via FMO-proteins and a chlorosome baseplate composed of CsmA proteins. Filamentous anoxygenic phototrophs of the phylum *Chloroflexi* lack the FMO complex, but instead use a protein complex called B808-866. Unlike the FMO proteins in green sulfur bacteria, B808-866 proteins are embedded in the cytoplasmic membrane and surround type-II reaction centers, providing the link between the reaction centers and the baseplate.

The composition of the chlorosomes is mostly bacteriochlorophyll (BChl) with small amounts of carotenoids and quinones surrounded by a galactolipid monolayer. In *Chlorobi*, chlorosome monolayers can contain up to eleven different proteins. The proteins of *Chlorobi* are the ones currently best understood in terms of structure and function. These proteins are named CsmA through CsmF, CsmH through CsmK, and CsmX. Other Csm proteins with different letter suffixes can be found in *Chloroflexi* and *Ca. Chloracidobacterium*.

Within the chlorosome, the thousands of BChl pigment molecules have the ability to self assemble with each other, meaning they do not interact with protein scaffolding complexes for assembly. These pigments self assemble in lamellar structures about 10-30 nm wide.

Organization of the Light Harvesting Pigments

Bacteriochlorophyll and carotenoids are two molecules responsible for harvesting light energy. Current models of the organization of bacteriochlorophyll and carotenoids (the main constituents) inside the chlorosomes have put them in a lamellar organization, where the long farnesol tails of the bacteriochlorophyll intermix with carotenoids and each other, forming a structure resembling a lipid multilayer.

Recently, another study has determined the organization of the bacteriochlorophyll molecules in green sulfur bacteria. Because they have been so difficult to study, the chlorosomes in green sulfur bacteria are the last class of light-harvesting complexes to be characterized structurally by scientists. Each individual chlorosome has a unique organization and this variability in composition had prevented scientists from using X-ray crystallography to characterize the internal structure. To get around this problem, the team used a combination of different experimental approaches. Genetic techniques to create a mutant bacterium with a more regular internal structure, cryo-electron microscopy to identify the larger distance constraints for the chlorosome, solid-state nuclear magnetic resonance (NMR) spectroscopy to determine the structure of the chlorosome's component chlorophyll molecules, and modeling to bring together all of the pieces and create a final picture of the chlorosome.

To create the mutant, three genes were inactivated that green sulfur bacteria acquired late in their evolution. In this way it was possible to go backward in evolutionary time to an intermediate state with much less variable and better ordered chlorosome organelles than the wild-type. The chlorosomes were isolated from the mutant and the wild-type forms of the bacteria. Cryo-electron microscopy was used to take pictures of the chlorosomes. The images reveal that the chlorophyll molecules inside chlorosomes have a nanotube shape. The team then used MAS NMR spectroscopy to resolve the microscopic arrangement of chlorophyll inside the chlorosome. With distance constraints and DFT ring current analyses the organization was found to consist of unique syn-anti monomer stacking. The combination of NMR, cryo-electron microscopy and modeling enabled the scientists to determine that the chlorophyll molecules in green sulfur bacteria are arranged in helices. In the mutant bacteria, the chlorophyll molecules are positioned at a nearly 90-degree angle in relation to the long axis of the nanotubes, whereas the angle is less steep in the wild-type organism. The structural framework can accommodate disorder to improve the biological light harvesting function, which implies that a less ordered structure has a better performance.

PHOTOSYNTHETIC EFFICIENCY

The photosynthetic efficiency is the fraction of light energy converted into chemical energy during photosynthesis in plants and algae. Photosynthesis can be described by the simplified chemical reaction:

$$6H_2O + 6CO_2 + energy \rightarrow C_6H_{12}O_6 + 6O_2$$

where $C_6H_{12}O_6$ is glucose (which is subsequently transformed into other sugars, cellulose, lignin, and so forth). The value of the photosynthetic efficiency is dependent on how light energy is defined – it depends on whether we count only the light that is absorbed, and on what kind of light is used. It takes eight (or perhaps 10 or more) photons to utilize one molecule of CO_2. The Gibbs free energy for converting a mole of CO_2 to glucose is 114 kcal, whereas eight moles of photons of wavelength 600 nm contains 381 kcal, giving a nominal efficiency of 30%. However, photosynthesis can occur with light up to wavelength 720 nm so long as there is also light at wavelengths below 680 nm to keep Photosystem II operating. Using longer wavelengths means less light energy is needed for the same number of photons and therefore for the same amount of photosynthesis. For actual sunlight, where only 45% of the light is in the photosynthetically active wavelength range, the theoretical maximum efficiency of solar energy conversion is approximately 11%. In actuality, however, plants do not absorb all incoming sunlight (due to reflection, respiration requirements of photosynthesis and the need for optimal solar radiation levels) and do not convert all harvested energy into biomass, which results in a maximum overall photosynthetic efficiency of 3 to 6% of total solar radiation. If photosynthesis is inefficient, excess light energy must be dissipated to avoid damaging the photosynthetic apparatus. Energy can be dissipated as heat (non-photochemical quenching), or emitted as chlorophyll fluorescence.

Typical Efficiencies

Plants

Quoted values sunlight-to-biomass efficiency:

Plant	Efficiency
Plants, typical	0.1% 0.2–2%3.5-4.3%
Typical crop plants	1–2%

The following is a breakdown of the energetics of the photosynthesis process from *Photosynthesis* by Hall and Rao:

Starting with the solar spectrum falling on a leaf, 47% lost due to photons outside the 400–700 nm active range (chlorophyll utilizes photons between 400 and 700 nm, extracting the energy of one 700 nm photon from each one) 30% of the in-band photons are lost due to incomplete absorption or photons hitting components other than chloroplasts 24% of the absorbed photon energy is lost due to degrading short wavelength photons to the 700 nm energy level 68% of the utilized energy is lost in conversion into d-glucose 35–45% of the glucose is consumed by the leaf in the processes of dark and photo respiration.

Stated another way: 100% sunlight → non-bioavailable photons waste is 47%, leaving 53% (in the 400–700 nm range) → 30% of photons are lost due to incomplete absorption, leaving 37% (absorbed photon energy) → 24% is lost due to wavelength-mismatch degradation to 700 nm energy, leaving 28.2% (sunlight energy collected by chlorophyll) → 32% efficient conversion of ATP and NADPH to d-glucose, leaving 9% (collected as sugar) → 35–40% of sugar is recycled/consumed by the leaf in dark and photo-respiration, leaving 5.4% net leaf efficiency.

Many plants lose much of the remaining energy on growing roots. Most crop plants store ~0.25% to 0.5% of the sunlight in the product (corn kernels, potato starch, etc.).

Photosynthesis increases linearly with light intensity at low intensity, but at higher intensity this is no longer the case. Above about 10,000 lux or ~100 watts/square meter the rate no longer increases. Thus, most plants can only utilize ~10% of full mid-day sunlight intensity. This dramatically reduces average achieved photosynthetic efficiency in fields compared to peak laboratory results. However, real plants (as opposed to laboratory test samples) have lots of redundant, randomly oriented leaves. This helps to keep the average illumination of each leaf well below the mid-day peak enabling the plant to achieve a result closer to the expected laboratory test results using limited illumination.

Only if the light intensity is above a plant specific value, called the compensation point the plant assimilates more carbon and releases more oxygen by photosynthesis than it consumes by cellular respiration for its own current energy demand. Photosynthesis measurement systems are not designed to directly measure the amount of light absorbed by the leaf. Nevertheless, the light response curves that the class produces do allow comparisons in photosynthetic efficiency between plants.

Algae and other Monocellular Organisms

From a 2010 study by the University of Maryland, photosynthesizing Cyanobacteria have been shown to be a significant species in the global carbon cycle, accounting for 20–30% of Earth's photosynthetic productivity and convert solar energy into biomass-stored chemical energy at the rate of ~450 TW.

Efficiencies of Various Biofuel Crops

Popular choices for plant biofuels include: oil palm, soybean, castor oil, sunflower oil, safflower oil, corn ethanol, and sugar cane ethanol.

An analysis of a proposed Hawaiian oil palm plantation claimed to yield 600 gallons of biodiesel per acre per year. That comes to 2835 watts per acre or 0.7 W/m^2. Typical insolation in Hawaii is around 5.5 kWh/(m^{2}day) or 230 W/m^2. For this particular oil palm plantation, if it delivered the claimed 600 gallons of biodiesel per acre per year, would be converting 0.3% of the incident solar energy to chemical fuel. Total photosynthetic efficiency would include more than just the biodiesel oil, so this 0.3% number is something of a lower bound.

Contrast this with a typical photovoltaic installation, which would produce an average of roughly 22 W/m^2 (roughly 10% of the average insolation), throughout the year. Furthermore, the photovoltaic panels would produce electricity, which is a high-quality form of energy, whereas converting the biodiesel into mechanical energy entails the loss of a large portion of the energy. On the other hand, a liquid fuel is much more convenient for a vehicle than electricity, which has to be stored in heavy, expensive batteries.

Most crop plants store ~0.25% to 0.5% of the sunlight in the product (corn kernels, potato starch, etc.), sugar cane is exceptional in several ways to yield peak storage efficiencies of ~8%.

Ethanol fuel in Brazil has a calculation that results in: "Per hectare per year, the biomass produced corresponds to 0.27 TJ. This is equivalent to 0.86 W/m^2. Assuming an average insolation of 225 W/m^2, the photosynthetic efficiency of sugar cane is 0.38%." Sucrose accounts for little more than 30% of the chemical energy stored in the mature plant; 35% is in the leaves and stem tips, which are left in the fields during harvest, and 35% are in the fibrous material (bagasse) left over from pressing.

C_3 vs. C_4 and CAM plants

C_3 plants use the Calvin cycle to fix carbon. C_4 plants use a modified Calvin cycle in which they separate Ribulose-1,5-bisphosphate carboxylase oxygenase (RuBisCO) from atmospheric oxygen, fixing carbon in their mesophyll cells and using oxaloacetate and malate to ferry the fixed carbon to RuBisCO and the rest of the Calvin cycle enzymes isolated in the bundle-sheath cells. The intermediate compounds both contain four carbon atoms, which gives C_4. In Crassulacean acid metabolism (CAM), time isolates functioning RuBisCo (and the other Calvin cycle enzymes) from high oxygen concentrations produced by photosynthesis, in that O_2 is evolved during the day, and allowed to dissipate then, while at night atmospheric CO_2 is taken up and stored as malic or other acids. During the day, CAM plants close stomata and use stored acids as carbon sources for sugar, etc. production.

The C_3 pathway requires 18 ATP and 12 NADPH for the synthesis of one molecule of glucose (3 ATP + 2 NADPH per CO_2 fixed) while the C_4 pathway requires 30 ATP and 12 NADPH (C_3 + 2 ATP per CO_2 fixed). In addition, we can take into account that each NADPH is equivalent to 3 ATP, that means both pathways require 36 additional (equivalent of) ATP. Despite this reduced ATP efficiency, C_4 is an evolutionary advancement, adapted to areas of high levels of light, where the reduced ATP efficiency is more than offset by the use of increased light. The ability to thrive despite restricted water availability maximizes the ability to use available light. The simpler C_3 cycle which operates in most plants is adapted to wetter darker environments, such as many northern latitudes.Corn, sugar cane, and sorghum are C_4 plants. These plants are economically important in part because of their relatively high photosynthetic efficiencies compared to many other crops. Pineapple is a CAM plant.

Photorespiration

One efficiency-focused research topic is improving the efficiency of photorespiration. Around 25 percent of the time RuBisCO incorrectly collects oxygen molecules instead of CO_2, creating CO_2 and ammonia that disrupt the photosynthesis process. Plants remove these byproducts via photorespiration, requiring energy and nutrients that would otherwise increase photosynthetic output. In C_3 plants photorespiration can consume 20-50% of photosynthetic energy. The research shortened photosynthetic pathways in tobacco. Engineered crops grew taller and faster, yielding up to 40 percent more biomass. The study employed synthetic biology to construct new metabolic pathways and assessed their efficiency with and without transporter RNAi. The most efficient pathway increased light-use efficiency by 17%.

PI CURVE

The PI (photosynthesis-irradiance) curve is a graphical representation of the empirical relationship between solar irradiance and photosynthesis. A derivation of the Michaelis–Menten curve, it shows the generally positive correlation between light intensity and photosynthetic rate. Plotted along the *x*-axis is the independent variable, light intensity (irradiance), while the *y*-axis is reserved for the dependent variable, photosynthetic rate.

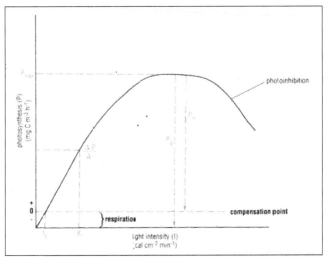

P v I curve

The PI curve can be applied to terrestrial and marine reactions but is most commonly used to explain ocean-dwelling phytoplankton's photosynthetic response to changes in light intensity. Using this tool to approximate biological productivity is important because phytoplankton contribute ~50% of total global carbon fixation and are important suppliers to the marine food web.

Within the scientific community, the curve can be referred to as the PI, PE or Light Response Curve. While individual researchers may have their own preferences, all are readily acceptable for use in the literature. Regardless of nomenclature, the photosynthetic rate in question can be described in terms of carbon (C) fixed per unit per time. Since individuals vary in size, it is also useful to normalise C concentration to Chlorophyll a (an important photosynthetic pigment) to account for specific biomass.

Equations

There are two simple derivations of the equation that are commonly used to generate the hyperbolic curve. The first assumes photosynthetic rate increases with increasing light intensity until Pmax is reached and continues to photosynthesise at the maximum rate thereafter.

$$P = P_{max}[I] / (KI + [I])$$

- P = photosynthetic rate at a given light intensity,

    ○ Commonly denoted in units such as (mg C m-3 h-1) or (µg C µg Chl-a-1 h-1),

- Pmax = the maximum potential photosynthetic rate per individual,

- [I] = a given light intensity,

  - Commonly denoted in units such as (μMol photons m-2 s-1 or (Watts m-2 h-1).

- KI = half-saturation constant; the light intensity at which the photosynthetic rate proceeds at ½ Pmax,

  - Units reflect those used for [I].

Both Pmax and the initial slope of the curve, $\Delta P/\Delta I$, are species-specific, and are influenced by a variety of factors, such as nutrient concentration, temperature and the physiological capabilities of the individual. Light intensity is influenced by latitudinal position and undergo daily and seasonal fluxes which will also affect the overall photosynthetic capacity of the individual. These three parameters are predictable and can be used to predetermine the general PI curve a population should follow.

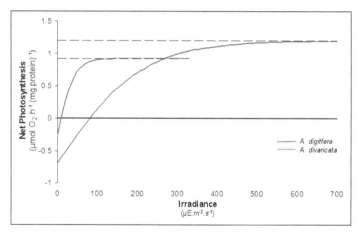

As can be seen in the graph, two species can have different responses to the same incremental changes in light intensity. Population A (in blue) has an initial rate higher than that of Population B (in red) and also exhibits a stronger rate change to increased light intensities at lower irradiance. Therefore, Population A will dominate in an environment with lower light availability. Although Population B has a slower photosynthetic response to increases in light intensity its Pmax is higher than that of Population A. This allows for eventual population dominance at greater light intensities. There are many determining factors influencing population success; using the PI curve to elicit predictions of rate flux to environmental changes is useful for monitoring phytoplankton bloom dynamics and ecosystem stability.

The second equation accounts for the phenomenon of photo inhibition. In the upper few meters of the ocean, phytoplankton may be subjected to irradiance levels that damage the chlorophyll-a pigment inside the cell, subsequently decreasing photosynthetic rate.

The response curve depicts photoinhibition as a decrease in photosynthetic rate at light intensities stronger than those necessary for achievement of Pmax.

$$P = P_{max}(1 - e^{-\alpha I/P_{max}})e^{-\beta I/P_{max}}$$

Terms not included in the above equation are:

- βI = light intensity at the start of photoinhibition.

- αI = a given light intensity.

Example of PI Curve

Data sets showing interspecific differences and population dynamics.

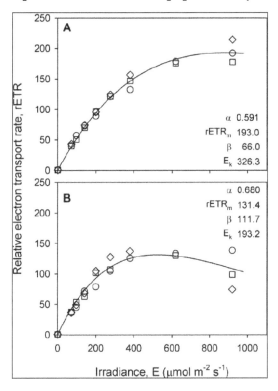

The hyperbolic response between photosynthesis and irradiance, depicted by the PI curve, is important for assessing phytoplankton population dynamics, which influence many aspects of the marine environment.

PHOTORESPIRATION

Photorespiration (also known as the oxidative photosynthetic carbon cycle, or C_2 photosynthesis) refers to a process in plant metabolism where the enzyme RuBisCO oxygenates RuBP, wasting some of the energy produced by photosynthesis. The desired reaction is the addition of carbon dioxide to RuBP (carboxylation), a key step in the Calvin–Benson cycle, however approximately 25% of reactions by RuBisCO instead add oxygen to RuBP (oxygenation), creating a product that cannot be used within the Calvin–Benson cycle. This process reduces the efficiency of photosynthesis, potentially reducing photosynthetic output by 25% in C_3 plants. Photorespiration involves a complex network of enzyme reactions that exchange metabolites between chloroplasts, leaf peroxisomes and mitochondria.

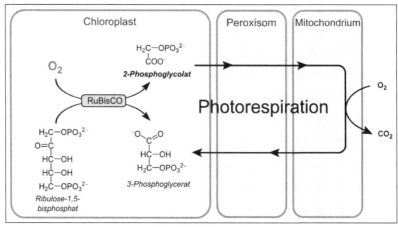

Simplified C_2 cycle

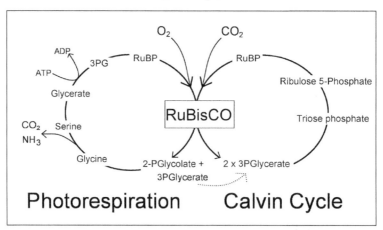

Simplified photorespiration and Calvin cycle.

The oxygenation reaction of RuBisCO is a wasteful process because 3-phosphoglycerate (G3P) is created at a reduced rate and higher metabolic cost compared with RuBP carboxylase activity. While photorespiratory carbon cycling results in the formation of G3P eventually, around 25% of carbon fixed by photorespiration is re-released as CO 2 and nitrogen, as ammonia. Ammonia must then be detoxified at a substantial cost to the cell. Photorespiration also incurs a direct cost of one ATP and one NAD(P)H.

While it is common to refer to the entire process as photorespiration, technically the term refers only to the metabolic network which acts to rescue the products of the oxygenation reaction (phosphoglycolate).

Photorespiratory Reactions

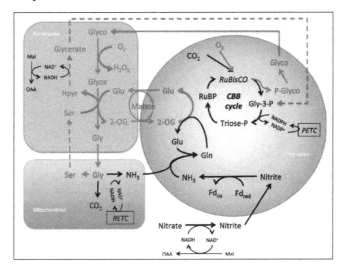

Addition of molecular oxygen to ribulose-1,5-bisphosphate produces 3-phosphoglycerate (PGA) and 2-phosphoglycolate (2PG, or PG). PGA is the normal product of carboxylation, and productively enters the Calvin cycle. Phosphoglycolate, however, inhibits certain enzymes involved in photosynthetic carbon fixation (hence is often said to be an 'inhibitor of photosynthesis'). It is also relatively difficult to recycle: in higher plants it is salvaged by a series of reactions in the peroxisome, mitochondria, and again in the peroxisome where it is converted into glycerate. Glycerate reenters the chloroplast and by the same transporter that exports glycolate. A cost of 1 ATP is associated with conversion to 3-phosphoglycerate (PGA) (Phosphorylation), within the chloroplast, which is then free to re-enter the Calvin cycle.

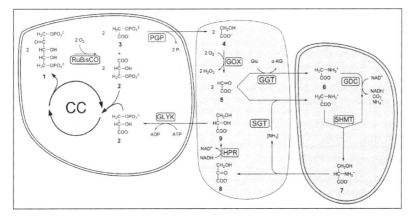

Several costs are associated with this metabolic pathway; the production of hydrogen peroxide in the peroxisome (associated with the conversion of glycolate to glyoxylate).

Hydrogen peroxide is a dangerously strong oxidant which must be immediately split into water and oxygen by the enzyme catalase. The conversion of 2×2 Carbon glycine to 1 C_3 serine in the mitochondria by the enzyme glycine-decarboxylase is a key step, which releases CO_2, NH_3, and reduces NAD to NADH. Thus, 1 CO_2 molecule is produced for every 2 molecules of O_2 (two deriving from RuBisCO and one from peroxisomal oxidations). The assimilation of NH_3 occurs via the GS-GOGAT cycle, at a cost of one ATP and one NADPH.

Cyanobacteria have three possible pathways through which they can metabolise 2-phosphoglycolate. They are unable to grow if all three pathways are knocked out, despite having a carbon concentrating mechanism that should dramatically reduce the rate of photorespiration.

Substrate Specificity of RuBisCO

The oxidative photosynthetic carbon cycle reaction is catalyzed by RuBP oxygenase activity:

$$RuBP + O_2 \rightarrow Phosphoglycolate + 3\text{-}phosphoglycerate + 2H^+$$

Oxygenase activity of RuBisCO.

During the catalysis by RuBisCO, an 'activated' intermediate is formed (an enediol intermediate) in the RuBisCO active site. This intermediate is able to react with either CO_2 or O_2. It has been demonstrated that the specific shape of the RuBisCO active site acts to encourage reactions with CO_2. Although there is a significant "failure" rate (~25% of reactions are oxygenation rather than carboxylation), this represents significant favouring of CO_2, when the relative abundance of the two gases is taken into account: in the current atmosphere, O_2 is approximately 500 times more abundant, and in solution O_2 is 25 times more abundant than CO_2.

The ability of RuBisCO to specify between the two gases is known as its selectivity factor (or Srel), and it varies between species, with angiosperms more efficient than other plants, but with little variation among the vascular plants.

A suggested explanation into RuBisCO's inability to discriminate completely between CO_2 and O_2 is that it is an evolutionary relic: The early atmosphere in which primitive plants originated contained very little oxygen, the early evolution of RuBisCO was not influenced by its ability to discriminate between O_2 and CO_2.

Conditions which Affect Photorespiration

Altered Substrate Availability: Lowered CO_2 or Increased O_2

Factors which influence this include the atmospheric abundance of the two gases, the supply of the gases to the site of fixation (i.e. in land plants: whether the stomata are open or closed), the length of the liquid phase (how far these gases have to diffuse through water in order to reach the reaction site). For example, when the stomata are closed to prevent water loss during drought: this limits the CO_2 supply, while O_2 production within the leaf will continue. In algae (and plants which photosynthesise underwater); gases have to diffuse significant distances through water, which results in a decrease in the availability of CO_2 relative to O_2. It has been predicted that the increase in ambient CO_2 concentrations predicted over the next 100 years may reduce the rate of photorespiration in most plants by around 50%.

Increased Temperature

At higher temperatures RuBisCO is less able to discriminate between CO_2 and O_2. This is because the enediol intermediate is less stable. Increasing temperatures also reduce the solubility of CO_2, thus reducing the concentration of CO_2 relative to O_2 in the chloroplast.

Biological Adaptation to Minimize Photorespiration

Certain species of plants or algae have mechanisms to reduce uptake of molecular oxygen by RuBisCO. These are commonly referred to as Carbon Concentrating Mechanisms (CCMs), as they increase the concentration of CO_2 so that RuBisCO is less likely to produce glycolate through reaction with O_2.

Biochemical Carbon Concentrating Mechanisms

Biochemical CCMs concentrate carbon dioxide in one temporal or spatial region, through metabolite exchange. C_4 and CAM photosynthesis both use the enzyme Phosphoenolpyruvate carboxylase (PEPC) to add CO_2 to a 3-Carbon sugar. PEPC is faster than RuBisCO, and more selective for CO_2.

C_4

C_4 plants capture carbon dioxide in their mesophyll cells (using an enzyme called phosphoenolpyruvate carboxylase which catalyzes the combination of carbon dioxide with a compound called phosphoenolpyruvate (PEP)), forming oxaloacetate. This oxaloacetate is then converted to malate and is transported into the bundle sheath cells (site of carbon dioxide fixation by RuBisCO) where oxygen concentration is low to avoid photorespiration. Here, carbon dioxide is removed from the malate and combined with RuBP by RuBisCO in the usual way, and the Calvin Cycle proceeds as normal. The CO_2 concentrations in the Bundle Sheath are approximately 10–20 fold higher than the concentration in the mesophyll cells.

Maize uses the C_4 pathway, minimizing photorespiration.

This ability to avoid photorespiration makes these plants more hardy than other plants in dry and hot environments, wherein stomata are closed and internal carbon dioxide levels are low. Under these conditions, photorespiration does occur in C_4 plants, but at a much reduced level compared with C_3 plants in the same conditions. C_4 plants include sugar cane, corn (maize), and sorghum.

CAM (Crassulacean Acid Metabolism)

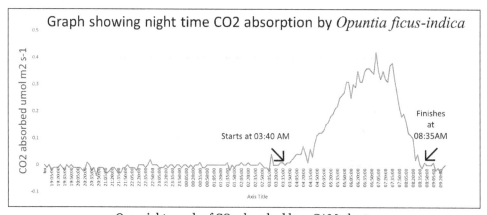

Overnight graph of CO_2 absorbed by a CAM plant.

CAM plants, such as cacti and succulent plants, also use the enzyme PEP carboxylase to capture carbon dioxide, but only at night. Crassulacean acid metabolism allows plants to conduct most of their gas exchange in the cooler night-time air, sequestering carbon in 4-carbon sugars which can be released to the photosynthesizing cells during the day. This allows CAM plants to reduce water loss (transpiration) by maintaining closed stomata during the day. CAM plants usually display other water-saving characteristics, such as thick cuticles, stomata with small apertures, and typically lose around 1/3 of the amount of water per CO_2 fixed.

Algae

There have been some reports of algae operating a biochemical CCM: shuttling metabolites within single cells to concentrate CO_2 in one area. This process is not fully understood.

Biophysical Carbon-concentrating Mechanisms

This type of carbon-concentrating mechanism (CCM) relies on a contained compartment within the cell into which CO_2 is shuttled, and where RuBisCO is highly expressed. In many species, biophysical CCMs are only induced under low carbon dioxide concentrations. Biophysical CCMs are more evolutionarily ancient than biochemical CCMs. There is some debate as to when biophysical CCMs first evolved, but it is likely to have been during a period of low carbon dioxide, after the Great Oxygenation Event (2.4 billion years ago). Low CO_2 periods occurred around 750, 650, and 320–270 million years ago.

Eukaryotic Algae

In nearly all species of eukaryotic algae (*Chloromonas* being one notable exception), upon induction of the CCM, ~95% of RuBisCO is densely packed into a single subcellular compartment: the pyrenoid. Carbon dioxide is concentrated in this compartment using a combination of CO_2 pumps, bicarbonate pumps, and carbonic anhydrases. The pyrenoid is not a membrane bound compartment, but is found within the chloroplast, often surrounded by a starch sheath (which is not thought to serve a function in the CCM).

Hornworts

Certain species of hornwort are the only land plants which are known to have a biophysical CCM involving concentration of carbon dioxide within pyrenoids in their chloroplasts.

Cyanobacteria

Cyanobacterial CCMs are similar in principle to those found in eukaryotic algae and hornworts, but the compartment into which carbon dioxide is concentrated has several structural differences. Instead of the pyrenoid, cyanobacteria contain carboxysomes, which have a protein shell, and linker proteins packing RuBisCO inside with a very regular structure. Cyanobacterial CCMs are much better understood than those found in eukaryotes, partly due to the ease of genetic manipulation of prokaryotes.

Possible Purpose of Photorespiration

Reducing photorespiration may not result in increased growth rates for plants. Photorespiration may be necessary for the assimilation of nitrate from soil. Thus, a reduction in photorespiration by genetic engineering or because of increasing atmospheric carbon dioxide (due to fossil fuel burning) may not benefit plants as has been proposed. Several physiological processes may be responsible for linking photorespiration and nitrogen assimilation. Photorespiration increases availability of NADH, which is required for the conversion of nitrate to nitrite. Certain nitrite transporters also transport bicarbonate, and elevated CO_2 has been shown to suppress nitrite transport into chloroplasts. However, in an agricultural setting, replacing the native photorespiration pathway with an engineered synthetic pathway to metabolize glycolate in the chloroplast resulted in a 40 percent increase in crop growth.

Although photorespiration is greatly reduced in C_4 species, it is still an essential pathway – mutants without functioning 2-phosphoglycolate metabolism cannot grow in normal conditions. One mutant was shown to rapidly accumulate glycolate.

Although the functions of photorespiration remain controversial, it is widely accepted that this pathway influences a wide range of processes from bioenergetics, photosystem II function, and carbon metabolism to nitrogen assimilation and respiration. The oxygenase reaction of RuBisCO may prevent CO_2 depletion near its active sites and contributes to the regulation of CO_2 concentration in the atmosphere. The photorespiratory pathway is a major source of hydrogen peroxide (H_2O_2) in photosynthetic cells. Through H_2O_2 production and pyrimidine nucleotide interactions, photorespiration makes a key contribution to cellular redox homeostasis. In so doing, it influences multiple signalling pathways, in particular, those that govern plant hormonal responses controlling growth, environmental and defense responses, and programmed cell death.

It has been postulated that photorespiration may function as a "safety valve", preventing the excess of reductive potential coming from an overreduced NADPH-pool from reacting with oxygen and producing free radicals, as these can damage the metabolic functions of the cell by subsequent oxidation of membrane lipids, proteins or nucleotides. The mutants deficient in photorespiratory enzymes are characterized by a high redox level in the cell, impaired stomatal regulation, and accumulation of formate.

CRASSULACEAN ACID METABOLISM

Crassulacean acid metabolism, also known as CAM photosynthesis, is a carbon fixation pathway that evolved in some plants as an adaptation to arid conditions. In a plant using full CAM, the stomata in the leaves remain shut during the day to reduce evapo-transpiration, but open at night to collect and allow carbon dioxide (CO_2) to diffuse into the mesophyll cells. The CO_2 is stored as the four-carbon acid malate in vacuoles at night, and then in the daytime, the malate is transported to chloroplasts where it is converted back to CO_2, which is then used during photosynthesis. The pre-collected CO_2 is concentrated around the enzyme RuBisCO, increasing photosynthetic efficiency. The mechanism was first discovered in plants of the family Crassulaceae.

The pineapple is an example of a CAM plant.

CAM is an adaptation for increased efficiency in the use of water, and so is typically found in plants growing in arid conditions.

During the Night

During the night, a plant employing CAM has its stomata open, allowing CO_2 to enter and be fixed as organic acids by a PEP reaction similar to the C_4 pathway. The resulting organic acids are stored in vacuoles for later use, as the Calvin cycle cannot operate without ATP and NADPH, products of light-dependent reactions that do not take place at night.

During the Day

During the day the stomata close to conserve water, and the CO_2-storing organic acids are released from the vacuoles of the mesophyll cells. An enzyme in the stroma of chloroplasts releases the CO_2, which enters into the Calvin cycle so that photosynthesis may take place.

Benefits

The most important benefit of CAM to the plant is the ability to leave most leaf stomata closed during the day. Plants employing CAM are most common in arid environments, where water comes at a premium. Being able to keep stomata closed during the hottest and driest part of the day reduces the loss of water through evapotranspiration, allowing such plants to grow in environments that would otherwise be far too dry. Plants using only C_3 carbon fixation, for example, lose 97% of the water they uptake through the roots to transpiration - a high cost avoided by plants able to employ CAM.

Comparison with C_4 Metabolism

CAM is named after the family Crassulaceae, to which the jade plant belongs.

The C_4 pathway bears resemblance to CAM; both act to concentrate CO_2 around RuBisCO, thereby increasing its efficiency. CAM concentrates it temporally, providing CO_2 during the day, and not at night, when respiration is the dominant reaction. C_4 plants, in contrast, concentrate CO_2 spatially, with a RuBisCO reaction centre in a "bundle sheath cell" being inundated with CO_2. Due to the inactivity required by the CAM mechanism, C_4 carbon fixation has a greater efficiency in terms of PGA synthesis.

Biochemistry

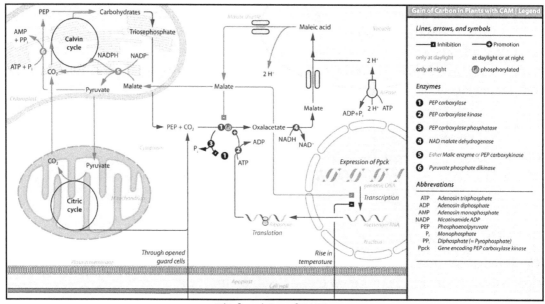

Biochemistry of CAM.

Plants with CAM must control storage of CO_2 and its reduction to branched carbohydrates in space and time.

At low temperatures (frequently at night), plants using CAM open their stomata, CO_2 molecules diffuse into the spongy mesophyll's intracellular spaces and then into the cytoplasm. Here, they can meet phosphoenolpyruvate (PEP), which is a phosphorylated triose. During this time, the plants are synthesizing a protein called PEP carboxylase kinase (PEP-C kinase), whose expression can be inhibited by high temperatures (frequently at daylight) and the presence of malate. PEP-C kinase phosphorylates its target enzyme PEP carboxylase (PEP-C). Phosphorylation dramatically enhances the enzyme's capability to catalyze the formation of oxaloacetate, which can be subsequently transformed into malate by NAD^+ malate dehydrogenase. Malate is then transported via malate shuttles into the vacuole, where it is converted into the storage form malic acid. In contrast to PEP-C kinase, PEP-C is synthesized all the time but almost inhibited at daylight either by dephosphorylation via PEP-C phosphatase or directly by binding malate. The latter is not possible at low temperatures, since malate is efficiently transported into the vacuole, whereas PEP-C kinase readily inverts dephosphorylation.

In daylight, plants using CAM close their guard cells and discharge malate that is subsequently transported into chloroplasts. There, depending on plant species, it is cleaved into pyruvate and CO_2 either by malic enzyme or by PEP carboxykinase. CO_2 is then introduced into the Calvin cycle, a coupled and self-recovering enzyme system, which is used to build branched carbohydrates. The by-product pyruvate can be further degraded in the mitochondrial citric acid cycle, thereby providing additional CO_2 molecules for the Calvin Cycle. Pyruvate can also be used to recover PEP via pyruvate phosphate dikinase, a high-energy step, which requires ATP and an additional phosphate. During the following cool night, PEP is finally exported into the cytoplasm, where it is involved in fixing carbon dioxide via malate.

Use by Plants

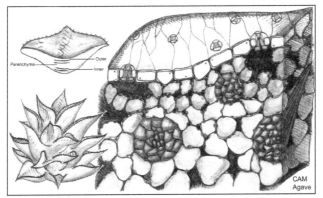

Cross section of a CAM (Crassulacean acid metabolism) plant,
specifically of an agave leaf. Vascular bundles shown.

Plants use CAM to different degrees. Some are "obligate CAM plants", i.e. they use only CAM in photosynthesis, although they vary in the amount of CO_2 they are able to store as organic acids; they are sometimes divided into "strong CAM" and "weak CAM" plants on this basis. Other plants show "inducible CAM", in which they are able to switch between using either the C_3 or C_4 mechanism and CAM depending on environmental conditions. Another group of plants employ "CAM-cycling", in which their stomata do not open at night; the plants instead recycle CO_2 produced by respiration as well as storing some CO_2 during the day.

Plants showing inducible CAM and CAM-cycling are typically found in conditions where periods of water shortage alternate with periods when water is freely available. Periodic drought – a feature of semi-arid regions – is one cause of water shortage. Plants which grow on trees or rocks (as epiphytes or lithophytes) also experience variations in water availability. Salinity, high light levels and nutrient availability are other factors which have been shown to induce CAM.

Since CAM is an adaptation to arid conditions, plants using CAM often display other xerophytic characters, such as thick, reduced leaves with a low surface-area-to-volume ratio; thick cuticle; and stomata sunken into pits. Some shed their leaves during the dry season; others (the succulents) store water in vacuoles. CAM also causes taste differences: plants may have an increasingly sour taste during the night yet become sweeter-tasting during the day. This is due to malic acid being stored in the vacuoles of the plants' cells during the night and then being used up during the day.

Aquatic CAM

CAM photosynthesis is also found in aquatic species in at least 4 genera, including: *Isoetes, Crassula, Littorella, Sagittaria*, and possibly *Vallisneria*, being found in a variety of species e.g. *Isoetes howellii, Crassula aquatica*.

These plants follow the same nocturnal acid accumulation and daytime deacidification as terrestrial CAM species. However, the reason for CAM in aquatic plants is not due to a lack of available water, but a limited supply of CO_2. CO_2 is limited due to slow diffusion in water, 10000x slower than in air. The problem is especially acute under acid pH, where the only inorganic carbon species present is CO_2, with no available bicarbonate or carbonate supply.

Aquatic CAM plants capture carbon at night when it is abundant due to a lack of competition from other photosynthetic organisms. This also results in lowered *photorespiration* due to less photosynthetically generated oxygen.

Aquatic CAM is most marked in the summer months when there is increased competition for CO_2, compared to the winter months. However, in the winter months CAM still has a significant role.

Ecological and Taxonomic Distribution of CAM-using Plants

The majority of plants possessing CAM are either epiphytes (e.g., orchids, bromeliads) or succulent xerophytes (e.g., cacti, cactoid *Euphorbia*s), but CAM is also found in hemiepiphytes (e.g., *Clusia*); lithophytes (e.g., *Sedum*, *Sempervivum*); terrestrial bromeliads; wetland plants (e.g., *Isoetes*, *Crassula* (*Tillaea*), *Lobelia*; and in one halophyte, *Mesembryanthemum crystallinum*; one non-succulent terrestrial plant, (*Dodonaea viscosa*) and one mangrove associate (*Sesuvium portulacastrum*).

Plants which are able to switch between different methods of carbon fixation include *Portulacaria afra*, better known as Dwarf Jade Plant, which normally uses C_3 fixation but can use CAM if it is drought-stressed, and *Portulaca oleracea*, better known as Purslane, which normally uses C_4 fixation but is also able to switch to CAM when drought-stressed.

CAM has evolved convergently many times. It occurs in 16,000 species (about 7% of plants), belonging to over 300 genera and around 40 families, but this is thought to be a considerable underestimate. It is found in quillworts (relatives of club mosses), in ferns, and in Gnetopsida, but the great majority of plants using CAM are angiosperms (flowering plants).

C_3 CARBON FIXATION

C_3 carbon fixation is the most common of three metabolic pathways for carbon fixation in photosynthesis, along with C_4 and CAM. This process converts carbon dioxide and ribulose bisphosphate (RuBP, a 5-carbon sugar) into two molecules of 3-phosphoglycerate through the following reaction:

$$CO_2 + H_2O + RuBP \rightarrow (2) \text{ 3-phosphoglycerate}$$

This reaction occurs in all plants as the first step of the Calvin–Benson cycle. (In C_4 and CAM plants, carbon dioxide is drawn out of malate and into this reaction rather than directly from the air.)

Plants that survive solely on C_3 fixation (C_3 plants) tend to thrive in areas where sunlight intensity is moderate, temperatures are moderate, carbon dioxide concentrations are around 200 ppm or higher, and groundwater is plentiful. The C_3 plants, originating during Mesozoic and Paleozoic eras, predate the C4 plants and still represent approximately 95% of Earth's plant biomass, including important food crops such as rice, wheat, soybeans and barley.

C_3 plants cannot grow in very hot areas because RuBisCO incorporates more oxygen into RuBP as temperatures increase. This leads to photorespiration (also known as the oxidative photosynthetic carbon cycle, or C_2 photosynthesis), which leads to a net loss of carbon and nitrogen from the plant and can therefore limit growth.

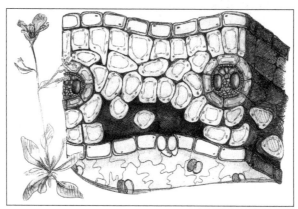

Cross section of a C_3 plant, specifically of an *Arabidopsis thaliana* leaf. Vascular bundles shown.

C_3 plants lose up to 97% of the water taken up through their roots to transpiration. In dry areas, C_3 plants shut their stomata to reduce water loss, but this stops CO_2 from entering the leaves and therefore reduces the concentration of CO_2 in the leaves. This lowers the CO_2:O_2 ratio and therefore also increases photorespiration. C_4 and CAM plants have adaptations that allow them to survive in hot and dry areas, and they can therefore out-compete C_3 plants in these areas.

The isotopic signature of C_3 plants shows higher degree of ^{13}C depletion than the C_4 plants, due to variation in fractionation of carbon isotopes in oxygenic photosynthesis across plant types.

Scientists have designed new metabolism pathways which reduces the losses to photorespiration, by more efficiently metabolizing the toxic glycolate produced. This resulted in over 40% increase in biomass production in their model organism (the tobacco plant) in their test conditions. The scientists are optimistic that this optimization can also be implemented in other C_3 crops like wheat.

C_4 CARBON FIXATION

C_4 carbon fixation or the Hatch–Slack pathway is a photosynthetic process in some plants. It is the first step in extracting carbon from carbon dioxide to be able to use it in sugar and other biomolecules. It is one of three known processes for carbon fixation. "C_4" refers to the four-carbon molecule that is the first product of this type of carbon fixation.

C_4 fixation is an elaboration of the more common C_3 carbon fixation and is believed to have evolved more recently. C_4 overcomes the tendency of the enzyme RuBisCO to wastefully fix oxygen rather than carbon dioxide in the process of photorespiration. This is achieved by ensuring that RuBisCO works in an environment where there is a lot of carbon dioxide and very little oxygen. CO_2 is shuttled via malate or aspartate from mesophyll cells to bundle-sheath cells. In these bundle-sheath cells CO_2 is released by decarboxylation of the malate. C_4 plants use PEP carboxylase to capture more CO_2 in the mesophyll cells. PEP (phosphoenolpyruvate, three carbons) binds to CO_2 to make oxaloacetic acid (OAA). OAA then makes malate (four carbons). Malate enters bundle sheath cells through plasmodesmata and releases the CO_2. These additional steps, however, require more energy in the form of ATP. Using this extra energy, C_4 plants are able to more efficiently fix carbon in drought, high temperatures, and limitations of nitrogen or CO_2. Since the more common C_3 pathway does not require this extra energy, it is more efficient in the other conditions.

The naming *Hatch–Slack pathway* is in honor of Marshall Davidson Hatch and C. R. Slack, who elucidated it in Australia in 1966.

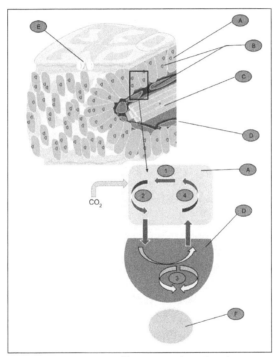

Kranz anatomy allows C_4 carbon fixation.

A: Mesophyll cell, B: Chloroplast, C: Vascular tissue, D: Bundle sheath cell, E: Stoma, F: Vascular tissue provides continuous source of water.

1. Carbon is fixed to produce oxaloacetate by PEP carboxylase.

2. The four-carbon molecule then exits the cell and enters the chloroplasts of bundle sheath cells.

3. It is then broken down releasing carbon dioxide and producing pyruvate. Carbon dioxide combines with ribulose 1,5 bisphosphate and proceeds to the Calvin cycle.

4. Pyruvate reenters the mesophyll cell. It then reacts with ATP to produce the beginning compound of the C_4 cycle.

C_4 Pathway

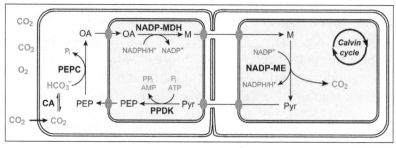

NADP-ME type C_4 pathway.

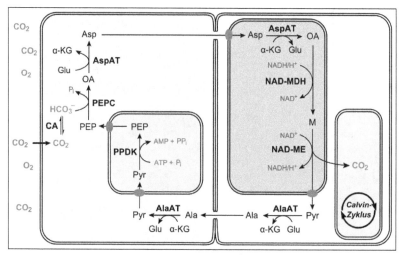

NAD-ME type C_4 pathway.

The first experiments indicating that some plants do not use C_3 carbon fixation but instead produce malate and aspartate in the first step of carbon fixation were done in the 1950s and early 1960s by Hugo P. Kortschak and Yuri Karpilov. The C_4 pathway was elucidated by Marshall Davidson Hatch and C. R. Slack, in Australia, in 1966; it is sometimes called the Hatch-Slack pathway.

In C_3 plants, the first step in the light-independent reactions of photosynthesis involves the fixation of CO_2 by the enzyme RuBisCO into 3-phosphoglycerate. However, due to the dual carboxylase and oxygenase activity of RuBisCo, some part of the substrate is oxidized rather than carboxylated, resulting in loss of substrate and consumption of energy, in what is known as photorespiration.

In order to bypass the photorespiration pathway, C_4 plants have developed a mechanism to efficiently deliver CO_2 to the RuBisCO enzyme. They use their specific leaf anatomy where chloroplasts exist not only in the mesophyll cells in the outer part of their leaves but in the bundle sheath cells as well. Instead of direct fixation to RuBisCO in the Calvin cycle, CO_2 is incorporated into a four-carbon organic acid, which has the ability to regenerate CO_2 in the chloroplasts of the bundle sheath cells. Bundle sheath cells can then use this CO_2 to generate carbohydrates by the conventional C_3 pathway.

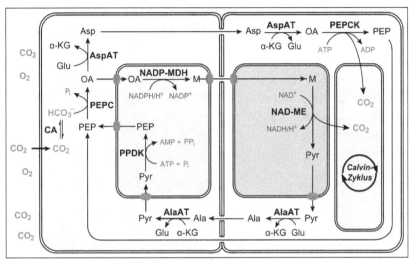

PEPCK type C_4 pathway

The first step in the pathway is the conversion of pyruvate to phosphoenolpyruvate (PEP), by the enzyme pyruvate orthophosphate dikinase. This reaction requires inorganic phosphate and ATP plus pyruvate, producing phosphoenolpyruvate, AMP, and inorganic pyrophosphate (PP_i). The next step is the fixation of CO_2 into oxaloacetate by the enzyme PEP carboxylase. Both of these steps occur in the mesophyll cells:

pyruvate + P_i + ATP → PEP + AMP + PP_i

PEP + CO_2 → oxaloacetate

PEP carboxylase has a lower Km for HCO_3^- — and, hence, higher affinity — than RuBisCO. Furthermore, O_2 is a very poor substrate for this enzyme. Thus, at relatively low concentrations of CO_2, most CO_2 will be fixed by this pathway.

The product is usually converted to malate, a simple organic compound, which is transported to the bundle-sheath cells surrounding a nearby vein. Here, it is decarboxylated to produce CO_2 and pyruvate. The CO_2 now enters the Calvin cycle and the pyruvate is transported back to the mesophyll cell.

Since every CO_2 molecule has to be fixed twice, first by four-carbon organic acid and second by RuBisCO, the C_4 pathway uses more energy than the C_3 pathway. The C_3 pathway requires 18 molecules of ATP for the synthesis of one molecule of glucose, whereas the C_4 pathway requires 30 molecules of ATP. This energy debt is more than paid for by avoiding losing more than half of photosynthetic carbon in photorespiration as occurs in some tropical plants,making it an adaptive mechanism for minimizing the loss.

There are several variants of this pathway:

1. The four-carbon acid transported from mesophyll cells may be malate, as above, or aspartate.

2. The three-carbon acid transported back from bundle-sheath cells may be pyruvate, as above, or alanine.

3. The enzyme that catalyses decarboxylation in bundle-sheath cells differs. In maize and sugarcane, the enzyme is NADP-malic enzyme; in millet, it is NAD-malic enzyme; and, in *Panicum maximum*, it is PEP carboxykinase.

C_4 Kranz Leaf Anatomy

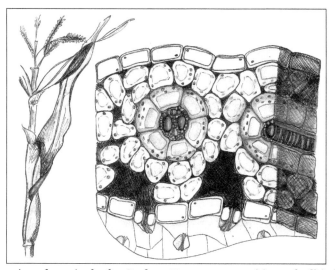

Cross section of a maize leaf, a C_4 plant. Kranz anatomy (rings of cells) shown.

The C_4 plants often possess a characteristic leaf anatomy called *kranz anatomy*, from the German word for wreath. Their vascular bundles are surrounded by two rings of cells; the inner ring, called bundle sheath cells, contains starch-rich chloroplasts lacking grana, which differ from those in mesophyll cells present as the outer ring. Hence, the chloroplasts are called dimorphic. The primary function of kranz anatomy is to provide a site in which CO_2 can be concentrated around RuBisCO, thereby avoiding photorespiration. In order to maintain a significantly higher CO_2 concentration in the bundle sheath compared to the mesophyll, the boundary layer of the kranz has a low conductance to CO_2, a property that may be enhanced by the presence of suberin. The carbon concentration mechanism in C_4 plants distinguishes their isotopic signature from other photosynthetic organisms.

Although most C_4 plants exhibit kranz anatomy, there are, however, a few species that operate a limited C_4 cycle without any distinct bundle sheath tissue. *Suaeda aralocaspica, Bienertia cycloptera, Bienertia sinuspersici* and *Bienertia kavirense* (all chenopods) are terrestrial plants that inhabit dry, salty depressions in the deserts of the Middle East. These plants have been shown to operate single-cell C_4 CO_2-concentrating mechanisms, which are unique among the known C_4 mechanisms. Although the cytology of both genera differs slightly, the basic principle is that fluid-filled vacuoles are employed to divide the cell into two separate areas. Carboxylation enzymes in the cytosol can, therefore, be kept separate from decarboxylase enzymes and RuBisCO in the chloroplasts, and a diffusive barrier can be established between the chloroplasts (which contain RuBisCO) and the cytosol. This enables a bundle-sheath-type area and a mesophyll-type area to be established within a single cell. Although this does allow a limited C_4 cycle to operate, it is relatively inefficient, with the occurrence of much leakage of CO_2 from around RuBisCO. There is also evidence for the exhibiting of inducible C_4 photosynthesis by non-kranz aquatic macrophyte *Hydrilla verticillata* under warm conditions, although the mechanism by which CO_2 leakage from around RuBisCO is minimised is currently uncertain.

Advantages of the C_4 Pathway

C_4 plants have a competitive advantage over plants possessing the more common C_3 carbon fixation pathway under conditions of drought, high temperatures, and nitrogen or CO_2 limitation. When grown in the same environment, at 30 °C, C_3 grasses lose approximately 833 molecules of water per CO_2 molecule that is fixed, whereas C_4 grasses lose only 277. This increased water use efficiency of C_4 grasses means that soil moisture is conserved, allowing them to grow for longer in arid environments.

C_4 carbon fixation has evolved on up to 61 independent occasions in 19 different families of plants, making it a prime example of convergent evolution. This convergence may have been facilitated by the fact that many potential evolutionary pathways to a C_4 phenotype exist, many of which involve initial evolutionary steps not directly related to photosynthesis. C_4 plants arose around 35 million years ago during the Oligocene (precisely when is difficult to determine) and did not become ecologically significant until around 6 to 7 million years ago, in the Miocene. C_4 metabolism in grasses originated when their habitat migrated from the shady forest undercanopy to more open environments, where the high sunlight gave it an advantage over the C_3 pathway. Drought was not necessary for its innovation; rather, the increased resistance to water stress was a byproduct of the pathway and allowed C_4 plants to more readily colonize arid environments.

Today, C_4 plants represent about 5% of Earth's plant biomass and 3% of its known plant species. Despite this scarcity, they account for about 23% of terrestrial carbon fixation. Increasing the proportion of C_4 plants on earth could assist biosequestration of CO_2 and represent an important climate change avoidance strategy. Present-day C_4 plants are concentrated in the tropics and subtropics (below latitudes of 45 degrees) where the high air temperature contributes to higher possible levels of oxygenase activity by RuBisCO, which increases rates of photorespiration in C_3 plants.

Plants that use C_4 Carbon Fixation

About 8,100 plant species use C_4 carbon fixation, which represents about 3% of all terrestrial species of plants. All these 8,100 species are angiosperms. C_4 carbon fixation is more common in monocots compared with dicots, with 40% of monocots using the C_4 pathway, compared with only 4.5% of dicots. Despite this, only three families of monocots use C_4 carbon fixation compared to 15 dicot families. Of the monocot clades containing C_4 plants, the grass (Poaceae) species use the C_4 photosynthetic pathway most. 46% of grasses are C_4 and together account for 61% of C_4 species. These include the food crops maize, sugar cane, millet, and sorghum. Of the dicot clades containing C_4 species, the order Caryophyllales contains the most species. Of the families in the Caryophyllales, the Chenopodiaceae use C_4 carbon fixation the most, with 550 out of 1,400 species using it. About 250 of the 1,000 species of the related Amaranthaceae also use C_4.

Maize (or corn) is a common C_4 plant.

Members of the sedge family Cyperaceae, and members of numerous families of eudicots – including Asteraceae (the daisy family), Brassicaceae (the cabbage family), and Euphorbiaceae (the spurge family) – also use C_4.

There are very few trees which use C_4. Only a handful are known: *Paulownia*, seven Hawaiian Euphorbia species and a few desert shrubs that reach the size and shape of trees with age.

Converting C_3 Plants to C_4

Given the advantages of C_4, a group of scientists from institutions around the world are working on the C_4 Rice Project to produce a strain of rice, naturally a C_3 plant, that uses the C_4 pathway by studying the C_4 plants maize and *Brachypodium*. As rice is the world's most important human food—it is the staple food for more than half the planet—having rice that is more efficient at converting sunlight into grain could have significant global benefits towards improving food security. The team claim C_4 rice could produce up to 50% more grain—and be able to do it with less water and nutrients.

The researchers have already identified genes needed for C_4 photosynthesis in rice and are now looking towards developing a prototype C_4 rice plant.

PHOTOSYNTHETICALLY ACTIVE RADIATION

Photosynthetically active radiation, often abbreviated PAR, designates the spectral range (wave band) of solar radiation from 400 to 700 nanometers that photosynthetic organisms are able to use in the process of photosynthesis. This spectral region corresponds more or less with the range of light visible to the human eye. Photons at shorter wavelengths tend to be so energetic that they can be damaging to cells and tissues, but are mostly filtered out by the ozone layer in the stratosphere. Photons at longer wavelengths do not carry enough energy to allow photosynthesis to take place.

Other living organisms, such as cyanobacteria, purple bacteria, and heliobacteria, can exploit solar light in slightly extended spectral regions, such as the near-infrared. These bacteria live in environments such as the bottom of stagnant ponds, sediment and ocean depths. Because of their pigments, they form colorful mats of green, red and purple.

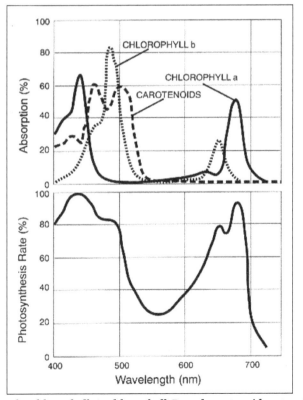

Top: Absorption spectra for chlorophyll-A, chlorophyll-B, and carotenoids extracted in a solution. Bottom: PAR action spectrum (oxygen evolution per incident photon) of an isolated chloroplast.

Chlorophyll, the most abundant plant pigment, is most efficient in capturing red and blue light. Accessory pigments such as carotenes and xanthophylls harvest some green light and pass it on to the photosynthetic process, but enough of the green wavelengths are reflected to give leaves their characteristic color. An exception to the predominance of chlorophyll is autumn, when chlorophyll is degraded (because it contains N and Mg) but the accessory pigments are not (because they only contain C, H and O) and remain in the leaf producing red, yellow and orange leaves.

In land plants, leaves absorb mostly red and blue light in the first layer of photosynthetic cells because of Chlorophyll absorbance. Green light, however, penetrates deeper into the leaf interior and can drive photosynthesis more efficiently than red light. Because green and yellow wavelengths can transmit through chlorophyll and the entire leaf itself, they play a crucial role in growth beneath the plant canopy.

PAR measurement is used in agriculture, forestry and oceanography. One of the requirements for productive farmland is adequate PAR, so PAR is used to evaluate agricultural investment potential. PAR sensors stationed at various levels of the forest canopy measure the pattern of PAR availability and utilization. Photosynthetic rate and related parameters can be measured non-destructively using a photosynthesis system, and these instruments measure PAR and sometimes control PAR at set intensities. PAR measurements are also used to calculate the euphotic depth in the ocean.

In these contexts, the reason PAR is preferred over other lighting metrics such as luminous flux and illuminance is that these measures are based on human perception of brightness, which is strongly green biased and does not accurately describe the quantity of light usable for photosynthesis.

Units

The irradiance of PAR can be measured in energy units (W/m^2), which is relevant in energy-balance considerations for photosynthetic organisms.

However, photosynthesis is a quantum process and the chemical reactions of photosynthesis are more dependent on the number of photons than the energy contained in the photons. Therefore, plant biologists often quantify PAR using the number of photons in the 400-700 nm range received by a surface for a specified amount of time, or the Photosynthetic Photon Flux Density (PPFD). PPFD is normally measured using mol $m^{-2}s^{-1}$. In relation to plant growth and morphology, it is better to characterise the light availability for plants by means of the Daily Light Integral (DLI), which is the daily flux of photons per ground area, and includes both diurnal variation as well as variation in day length.

In the past PPFD was often expressed using einstein units, i.e., $\mu E\ m^{-2}s^{-1}$. An einstein is simply a mole of photons.

Yield Photon Flux

There are two common measures of photosynthetically active radiation: photosynthetic photon flux (PPF) and yield photon flux (YPF). PPF values all photons from 400 to 700 nm equally, while YPF weights photons in the range from 360 to 760 nm based on a plant's photosynthetic response.

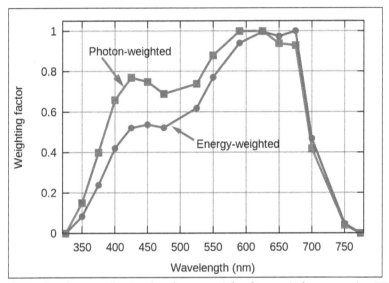

Weighting factor for photosynthesis. The photon-weighted curve is for converting PPF to YPF; the energy-weighted curve is for weighting PAR expressed in watts or joules.

PAR as described with PPF does not distinguish between different wavelengths between 400 and 700 nm, and assumes that wavelengths outside this range have zero photosynthetic action. If the exact spectrum of the light is known, the photosynthetic photon flux density (PPFD) values in μmol s^{-1}m^{-2}) can be modified by applying different weighting factors to different wavelengths. This results in a quantity called the yield photon flux (YPF). The red curve in the graph shows that photons around 610 nm (orange-red) have the highest amount of photosynthesis per photon. However, because short-wavelength photons carry more energy per photon, the maximum amount of photosynthesis per incident unit of energy is at a longer wavelength, around 650 nm (deep red).

It has been noted that there is considerable misunderstanding over the effect of light quality on plant growth. Many manufacturers claim significantly increased plant growth due to light quality (high YPF). The YPF curve indicates that orange and red photons between 600 and 630 nm can result in 20 to 30% more photosynthesis than blue or cyan photons between 400 and 540 nm. But the YPF curve was developed from short-term measurements made on single leaves in low light. More recent longer-term studies with whole plants in higher light indicate that light quality may have a smaller effect on plant growth rate than light quantity. Blue light, while not delivering as many photons per joule, encourages leaf growth and affects other outcomes.

The conversion between energy-based PAR and photon-based PAR depends on the spectrum of the light source. The following table shows the conversion factors from watts for black-body spectra that are truncated to the range 400–700 nm. It also shows the luminous efficacy for these light sources and the fraction of a real black-body radiator that is emitted as PAR.

T (K)	η_v (lm/W)	η_{photon} (μmol/J or μmol s^{-1}W^{-1})	η_{photon} (mol day^{-1} W^{-1})	η_{PAR} (W/W)
3000 (warm white)	269	4.98	0.43	0.0809
4000	277	4.78	0.413	0.208
5800 (daylight)	265	4.56	0.394	0.368

For example, a light source of 1000 lm at a color temperature of 5800 K would emit approximately $1000/265 = 3.8$ W of PAR, which is equivalent to $3.8 \times 4.56 = 17.3$ μmol/s. For a black-body light source at 5800 K, such as the sun is approximately, a fraction 0.368 of its total emitted radiation is emitted as PAR. For artificial light sources, that usually do not have a black-body spectrum, these conversion factors are only approximate.

The quantities in the table are calculated as:

$$\eta_v(T) = \frac{\int_{\lambda_1}^{\lambda_2} B(\lambda,T)683\,[\text{lm/W}]y(\lambda)d\lambda}{\int_{\lambda_1}^{\lambda_2} B(\lambda,T)d\lambda},$$

$$\eta_{photon}(T) = \frac{\int_{\lambda_1}^{\lambda_2} B(\lambda,T)\dfrac{\lambda}{hcN_A}d\lambda}{\int_{\lambda_1}^{\lambda_2} B(\lambda,T)d\lambda},$$

$$\eta_{PAR}(T) = \frac{\int_{\lambda_1}^{\lambda_2} B(\lambda,T)d\lambda}{\int_0^\infty B(\lambda,T)d\lambda},$$

where $B(\lambda,T)$ is the black-body spectrum according to Planck's law, y is the standard luminosity function, λ_1, λ_2 represent the wavelength range (400 700 nm) of PAR, and N_A is the Avogadro constant.

Second Law PAR Efficiency

Besides the amount of radiation reaching a plant in the PAR region of the spectrum, it is also important to consider the quality of such radiation. Radiation reaching a plant contains entropy as well as energy, and combining those two concepts the exergy can be determined. This sort of analysis is known as exergy analysis or second law analysis, and the exergy represents a measure of the useful work, i.e., the useful part of radiation which can be transformed into other forms of energy.

The spectral distribution of the exergy of radiation is defined as:

$$Ex_\lambda = L_\lambda(T) - L_\lambda(T_0) - T_0[S_\lambda(T) - S_\lambda(T_0)]$$

One of the advantages of working with the exergy is that it depends on the temperature of the emitter (the Sun), T, but also on the temperature of the receiving body (the plant), T_0, i.e., it includes the fact that the plant is emitting radiation. Naming $x = \dfrac{hc}{\lambda kT}$ and $y = \dfrac{hc}{\lambda kT_0}$, the exergy emissive power of radiation in a region is determined as:

$$\int_0^{\lambda_i} Ex(\lambda,T)d\lambda = \mathfrak{I}_{Ex_{0\to\lambda_i}} = \frac{15}{\pi^4}\sigma\Big\{T^3\Big[(T-T_0)x^3 Li_1(e^{-x}) + (3T - 4T_0)x^2 Li_2(e^{-x})$$

$$+ (6T - 8T_0)xLi_3(e^{-x}) + (6T - 8T_0)Li_4(e^{-x})\Big]$$

$$+ T_0^4\Big[y^2 Li_2(e^{-y}) + 2yLi_3(e^{-y}) + 2Li_4(e^{-y})\Big]\Big\}$$

Where $Li_s(z)$ is a special function called Polylogarithm. By definition, the exergy obtained by the receiving body is always lower than the energy radiated by the emitting blackbody, as a consequence of the entropy content in radiation. Thus, as a consequence of the entropy content, not all the radiation reaching the Earth's surface is "useful" to produce work. Therefore, the efficiency of a process involving radiation should be measured against its exergy, not its energy.

Using the expression above, the optimal efficiency or second law efficiency for the conversion of radiation to work in the PAR region (from $\lambda_1 = 400$ nm to $\lambda_2 = 700$ nm), for a blackbody at $T = 5800$ K and an organism at $T_0 = 300$ K is determined as:

$$\eta_{PAR}^{ex}(T) = \frac{\int_{\lambda_1}^{\lambda_2} Ex(\lambda,T)d\lambda}{\int_0^\infty L(\lambda,T)d\lambda} = 0.337563$$

about 8.3% lower than the value considered until now, as a direct consequence of the fact that the organisms which are using solar radiation are also emitting radiation as a consequence of their own temperature. Therefore, the conversion factor of the organism will be different depending on its temperature, and the exergy concept is more suitable than the energy one.

Measurement of PAR

Researchers at Utah State University compared measurements for PPF and YPF using different types of equipment. They measured the PPF and YPF of seven common radiation sources with a spectroradiometer, then compared with measurements from six quantum sensors designed to measure PPF, and three quantum sensors designed to measure YPF.

They found that the PPF and YPF sensors were the least accurate for narrow-band sources (narrow spectrum of light) and most accurate broad-band sources (fuller spectra of light). They found that PPF sensors were significantly more accurate under metal halide, low-pressure sodium and high-pressure sodium lamps than YPF sensors (>9% difference). Both YPF and PPF sensors were very inaccurate (>18% error) when used to measure light from red-light-emitting diodes.

ARTIFICIAL PHOTOSYNTHESIS

A sample of a photoelectric cell in a lab environment. Catalysts are added to the cell, which is submerged in water and illuminated by simulated sunlight. The bubbles seen are oxygen (forming on the front of the cell) and hydrogen (forming on the back of the cell).

Artificial photosynthesis is a chemical process that biomimics the natural process of photosynthesis to convert sunlight, water, and carbon dioxide into carbohydrates and oxygen. The term artificial photosynthesis is commonly used to refer to any scheme for capturing and storing the energy from sunlight in the chemical bonds of a fuel (a solar fuel). Photocatalytic water splitting converts water into hydrogen and oxygen and is a major research topic of artificial photosynthesis. Light-driven carbon dioxide reduction is another process studied that replicates natural carbon fixation.

Research of this topic includes the design and assembly of devices for the direct production of solar fuels, photoelectrochemistry and its application in fuel cells, and the engineering of enzymes and photoautotrophic microorganisms for microbial biofuel and biohydrogen production from sunlight.

The photosynthetic reaction can be divided into two half-reactions of oxidation and reduction, both of which are essential to producing fuel. In plant photosynthesis, water molecules are photo-oxidized to release oxygen and protons. The second phase of plant photosynthesis (also known as the Calvin-Benson cycle) is a light-independent reaction that converts carbon dioxide into glucose (fuel). Researchers of artificial photosynthesis are developing photocatalysts that are able to perform both of these reactions. Furthermore, the protons resulting from water splitting can be used for hydrogen production. These catalysts must be able to react quickly and absorb a large percentage of the incident solar photons.

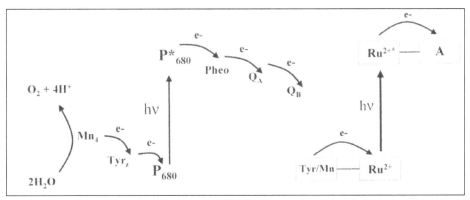

Natural (left) versus artificial photosynthesis (right).

Whereas photovoltaics can provide energy directly from sunlight, the inefficiency of fuel production from photovoltaic electricity (indirect process) and the fact that sunshine is not constant throughout the day sets a limit to its use. One way of using natural photosynthesis is for the production of a biofuel, which is an indirect process that suffers from low energy conversion efficiency (due to photosynthesis' own low efficiency in converting sunlight to biomass), the cost of harvesting and transporting the fuel, and conflicts due to the increasing need of land mass for food production. The purpose of artificial photosynthesis is to produce a fuel from sunlight that can be stored conveniently and used when sunlight is not available, by using direct processes, that is, to produce a solar fuel. With the development of catalysts able to reproduce the major parts of photosynthesis, water and sunlight would ultimately be the only needed sources for clean energy production. The only by-product would be oxygen, and production of a solar fuel has the potential to be cheaper than gasoline.

One process for the creation of a clean and affordable energy supply is the development of photocatalytic water splitting under solar light. This method of sustainable hydrogen production is a major objective for the development of alternative energy systems. It is also predicted to be one of the more, if not the most, efficient ways of obtaining hydrogen from water. The conversion of solar energy into hydrogen via a water-splitting process assisted by photo semiconductor catalysts is one of the most promising technologies in development. This process has the potential for large quantities of hydrogen to be generated in an ecologically sound manner. The conversion of solar energy into a clean fuel (H_2) under ambient conditions is one of the greatest challenges facing scientists in the twenty-first century.

Two methods are generally recognized for the construction of solar fuel cells for hydrogen production:

- A homogeneous system is one such that catalysts are not compartmentalized, that is, components are present in the same compartment. This means that hydrogen and oxygen are produced in the same location. This can be a drawback, since they compose an explosive mixture, demanding gas product separation. Also, all components must be active in approximately the same conditions (e.g., pH).

- A heterogeneous system has two separate electrodes, an anode and a cathode, making possible the separation of oxygen and hydrogen production. Furthermore, different components do not necessarily need to work in the same conditions. However, the increased complexity of these systems makes them harder to develop and more expensive.

Another area of research within artificial photosynthesis is the selection and manipulation of photosynthetic microorganisms, namely green microalgae and cyanobacteria, for the production of solar fuels. Many strains are able to produce hydrogen naturally, and scientists are working to improve them. Algae biofuels such as butanol and methanol are produced both at laboratory and commercial scales. This method has benefited from the development of synthetic biology, which is also being explored by the J. Craig Venter Institute to produce a synthetic organism capable of biofuel production. In 2017, an efficient process was developed to produce acetic acid from carbon dioxide using "cyborg bacteria".

Artificial photosynthesis was first anticipated by the Italian chemist Giacomo Ciamician during 1912. He proposed a switch from the use of fossil fuels to radiant energy provided by the sun and captured by technical photochemistry devices. In this switch he saw a possibility to lessen the difference between the rich north of Europe and poor south and ventured a guess that this switch from coal to solar energy would "not be harmful to the progress and to human happiness."

During the late 1960s, Akira Fujishima discovered the photocatalytic properties of titanium dioxide, the so-called Honda-Fujishima effect, which could be used for hydrolysis.

The Swedish Consortium for Artificial Photosynthesis, the first of its kind, was established during 1994 as a collaboration between groups of three different universities, Lund, Uppsala and Stockholm, being presently active around Lund and the Ångström Laboratories in Uppsala. The consortium was built with a multidisciplinary approach to focus on learning from natural photosynthesis and applying this knowledge in biomimetic systems.

Research of artificial photosynthesis is experiencing a boom at the beginning of the 21st century. During 2000, Commonwealth Scientific and Industrial Research Organisation (CSIRO) researchers publicized their intent to emphasize carbon dioxide capture and its conversion to hydrocarbons. In 2003, the Brookhaven National Laboratory announced the discovery of an important intermediate part of the reduction of CO_2 to CO (the simplest possible carbon dioxide reduction reaction), which could result in better catalysts.

Visible light water splitting with a one piece multijunction semiconductor cell (vs. UV light with titanium dioxide semiconductors) was first demonstrated and patented by William Ayers at Energy Conversion Devices during 1983. This group demonstrated water photolysis into hydrogen and oxygen, now referred to as an "artificial leaf" or "wireless solar water splitting" with a low cost, thin film amorphous silicon multijunction cell immersed directly in water. Hydrogen evolved on the front amorphous silicon surface decorated with various catalysts while oxygen evolved from the back metal substrate which also eliminated the hazard of mixed hydrogen/oxygen gas evolution. A Nafion membrane above the immersed cell provided a path for proton transport. The higher photovoltage available from the multijuction thin film cell with visible light was a major advance over previous photolysis attempts with UV sensitive single junction cells. The group's patent also lists several other semiconductor multijunction compositions in addition to amorphous silicon.

One of the disadvantages of artificial systems for water-splitting catalysts is their general reliance on scarce, expensive elements, such as ruthenium or rhenium. During 2008, with the funding of the United States Air Force Office of Scientific Research, MIT chemist and director of the Solar Revolution Project Daniel G. Nocera and postdoctoral fellow Matthew Kanan attempted to circumvent this problem by using a catalyst containing the cheaper and more abundant elements cobalt and phosphate. The catalyst was able to split water into oxygen and protons using sunlight, and could potentially be coupled to a hydrogen gas producing catalyst such as platinum. Furthermore, while the catalyst broke down during catalysis, it could self-repair. This experimental catalyst design was considered a major improvement by many researchers.

Whereas CO is the prime reduction product of CO_2, more complex carbon compounds are usually desired. During 2008, Andrew B. Bocarsly reported the direct conversion of carbon dioxide and water to methanol using solar energy in a very efficient photochemical cell.

While Nocera and coworkers had accomplished water splitting to oxygen and protons, a light-driven process to produce hydrogen is desirable. During 2009, the Leibniz Institute for Catalysis reported inexpensive iron carbonyl complexes able to do just that. During the same year, researchers at the University of East Anglia also used iron carbonyl compounds to achieve photoelectrochemical hydrogen production with 60% efficiency, this time using a gold electrode covered with layers of indium phosphide to which the iron complexes were linked. Both of these processes used a molecular approach, where discrete nanoparticles are responsible for catalysis.

During 2009, F. del Valle and K. Domen showed the effect of the thermal treatment in a closed atmosphere using $Cd_{1-x}Zn_xS$ photocatalysts. $Cd_{1-x}Zn_xS$ solid solution reports high activity in hydrogen production from water splitting under sunlight irradiation. A mixed heterogeneous/molecular approach by researchers at the University of California, Santa Cruz, during 2010, using both nitrogen-doped and cadmium selenide quantum dots-sensitized titanium dioxide nanoparticles and nanowires, also yielded photoproduced hydrogen.

Artificial photosynthesis remained an academic field for many years. However, in the beginning of 2009, Mitsubishi Chemical Holdings was reported to be developing its own artificial photosynthesis research by using sunlight, water and carbon dioxide to "create the carbon building blocks from which resins, plastics and fibers can be synthesized." This was confirmed with the establishment of the KAITEKI Institute later that year, with carbon dioxide reduction through artificial photosynthesis as one of the main goals.

During 2010, the United States Department of Energy established, as one of its Energy Innovation Hubs, the Joint Center for Artificial Photosynthesis. The mission of JCAP is to find a cost-effective method to produce fuels using only sunlight, water, and carbon-dioxide as inputs. JCAP is managed by a team from Caltech, directed by Professor Nathan Lewis and brings together more than 120 scientists and engineers from California Institute of Technology and its main partner, Lawrence Berkeley National Laboratory. JCAP also draws on the expertise and capabilities of key partners from Stanford University, the University of California at Berkeley, UCSB, University of California, Irvine, and University of California at San Diego, and the Stanford Linear Accelerator. Additionally, JCAP serves as a central hub for other solar fuels research teams across the United States, including 20 DOE Energy Frontier Research Center. The program has a budget of $122M over five years, subject to Congressional appropriation.

Also during 2010, a team directed by professor David Wendell at the University of Cincinnati successfully demonstrated photosynthesis in an artificial construct consisting of enzymes suspended in a foam housing.

During 2011, Daniel Nocera and his research team announced the creation of the first practical artificial leaf. In a speech at the 241st National Meeting of the American

Chemical Society, Nocera described an advanced solar cell the size of a poker card capable of splitting water into oxygen and hydrogen, approximately ten times more efficient than natural photosynthesis. The cell is mostly made of inexpensive materials that are widely available, works under simple conditions, and shows increased stability over previous catalysts: in laboratory studies, the authors demonstrated that an artificial leaf prototype could operate continuously for at least forty-five hours without a drop in activity. In May 2012, Sun Catalytix, the startup based on Nocera's research, stated that it will not be scaling up the prototype as the device offers few savings over other ways to make hydrogen from sunlight. (Sun Catalytix ended up later pivoting away from solar fuel to develop batteries to store energy for the power grid instead, and Lockheed bought the company for an undisclosed amount in 2014) Leading experts in the field have supported a proposal for a Global Project on Artificial Photosynthesis as a combined energy security and climate change solution. Conferences on this theme have been held at Lord Howe Island during 2011, at Chicheley Hall in the UK in 2014 and at Canberra and Lord Howe island during 2016.

In energy terms, natural photosynthesis can be divided in three steps:

- Light-harvesting complexes in bacteria and plants capture photons and transduce them into electrons, injecting them into the photosynthetic chain.

- Proton-coupled electron transfer along several cofactors of the photosynthetic chain, causing local, spatial charge separation.

- Redox catalysis, which uses the aforementioned transferred electrons to oxidize water to dioxygen and protons; these protons can in some species be utilized for dihydrogen production.

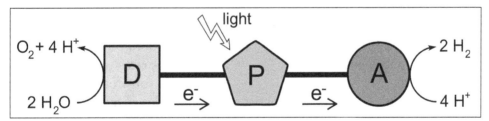

A triad assembly, with a photosensitizer (P) linked in tandem to a water oxidation catalyst (D) and a hydrogen evolving catalyst (A). Electrons flow from D to A when catalysis occurs.

Using biomimetic approaches, artificial photosynthesis tries to construct systems doing the same type of processes. Ideally, a triad assembly could oxidize water with one catalyst, reduce protons with another and have a photosensitizer molecule to power the whole system. One of the simplest designs is where the photosensitizer is linked in tandem between a water oxidation catalyst and a hydrogen evolving catalyst:

- The photosensitizer transfers electrons to the hydrogen catalyst when hit by light, becoming oxidized in the process.

- This drives the water splitting catalyst to donate electrons to the photosensitizer. In a triad assembly, such a catalyst is often referred to as a donor. The oxidized donor is able to perform water oxidation.

The state of the triad with one catalyst oxidized on one end and the second one reduced on the other end of the triad is referred to as a charge separation, and is a driving force for further electron transfer, and consequently catalysis, to occur. The different components may be assembled in diverse ways, such as supramolecular complexes, compartmentalized cells, or linearly, covalently linked molecules.

Research into finding catalysts that can convert water, carbon dioxide, and sunlight to carbohydrates or hydrogen is a current, active field. By studying the natural oxygen-evolving complex (OEC), researchers have developed catalysts such as the "blue dimer" to mimic its function or inorganic-based materials such as Birnessite with the similar building block as the OEC. Photoelectrochemical cells that reduce carbon dioxide into carbon monoxide (CO), formic acid (HCOOH) and methanol (CH_3OH) are under development. However, these catalysts are still very inefficient.

Hydrogen Catalysts

Hydrogen is the simplest solar fuel to synthesize, since it involves only the transference of two electrons to two protons. It must, however, be done stepwise, with formation of an intermediate hydride anion:

$$2\,e^- + 2\,H^+ \rightleftharpoons H^+ + H^- \rightleftharpoons H_2$$

The proton-to-hydrogen converting catalysts present in nature are hydrogenases. These are enzymes that can either reduce protons to molecular hydrogen or oxidize hydrogen to protons and electrons. Spectroscopic and crystallographic studies spanning several decades have resulted in a good understanding of both the structure and mechanism of hydrogenase catalysis. Using this information, several molecules mimicking the structure of the active site of both nickel-iron and iron-iron hydrogenases have been synthesized. Other catalysts are not structural mimics of hydrogenase but rather functional ones. Synthesized catalysts include structural H-cluster models, a dirhodium photocatalyst, and cobalt catalysts.

Water-oxidizing Catalysts

Water oxidation is a more complex chemical reaction than proton reduction. In nature, the oxygen-evolving complex performs this reaction by accumulating reducing equivalents (electrons) in a manganese-calcium cluster within photosystem II (PS II), then delivering them to water molecules, with the resulting production of molecular oxygen and protons:

$$2\,H_2O \rightarrow O_2 + 4\,H^+ + 4e^-$$

Without a catalyst (natural or artificial), this reaction is very endothermic, requiring high temperatures (at least 2500 K).

The exact structure of the oxygen-evolving complex has been hard to determine experimentally. As of 2011, the most detailed model was from a 1.9 Å resolution crystal structure of photosystem II. The complex is a cluster containing four manganese and one calcium ions, but the exact location and mechanism of water oxidation within the cluster is unknown. Nevertheless, bio-inspired manganese and manganese-calcium complexes have been synthesized, such as $[Mn_4O_4]$ cubane-type clusters, some with catalytic activity.

Some ruthenium complexes, such as the dinuclear μ-oxo-bridged "blue dimer" (the first of its kind to be synthesized), are capable of light-driven water oxidation, thanks to being able to form high valence states. In this case, the ruthenium complex acts as both photosensitizer and catalyst.

Many metal oxides have been found to have water oxidation catalytic activity, including ruthenium(IV) oxide (RuO_2), iridium(IV) oxide (IrO_2), cobalt oxides (including nickel-doped Co_3O_4), manganese oxide (including layered MnO_2 (birnessite), Mn_2O_3), and a mix of Mn_2O_3 with $CaMn_2O_4$. Oxides are easier to obtain than molecular catalysts, especially those from relatively abundant transition metals (cobalt and manganese), but suffer from low turnover frequency and slow electron transfer properties, and their mechanism of action is hard to decipher and, therefore, to adjust.

Recently Metal-Organic Framework (MOF)-based materials have been shown to be a highly promising candidate for water oxidation with first row transition metals. The stability and tunability of this system is projected to be highly beneficial for future development.

Photosensitizers

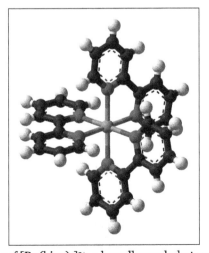

Structure of $[Ru(bipy)_3]^{2+}$, a broadly used photosensitizer.

Nature uses pigments, mainly chlorophylls, to absorb a broad part of the visible spectrum. Artificial systems can use either one type of pigment with a broad absorption range or combine several pigments for the same purpose.

Ruthenium polypyridine complexes, in particular tris(bipyridine)ruthenium(II) and its derivatives, have been extensively used in hydrogen photoproduction due to their efficient visible light absorption and long-lived consequent metal-to-ligand charge transfer excited state, which makes the complexes strong reducing agents. Other noble metal-containing complexes used include ones with platinum, rhodium and iridium.

Metal-free organic complexes have also been successfully employed as photosensitizers. Examples include eosin Y and rose bengal. Pyrrole rings such as porphyrins have also been used in coating nanomaterials or semiconductors for both homogeneous and heterogeneous catalysis.

As part of current research efforts artificial photonic antenna systems are being studied to determine efficient and sustainable ways to collect light for artificial photosynthesis. Gion Calzaferri (2009) describes one such antenna that uses zeolite L as a host for organic dyes, to mimic plant's light collecting systems. The antenna is fabricated by inserting dye molecules into the channels of zeolite L. The insertion process, which takes place under vacuum and at high temperature conditions, is made possible by the cooperative vibrational motion of the zeolite framework and of the dye molecules. The resulting material may be interfaced to an external device via a stopcock intermediate.

Carbon Dioxide Reduction Catalysts

In nature, carbon fixation is done by green plants using the enzyme RuBisCO as a part of the Calvin cycle. RuBisCO is a rather slow catalyst compared to the vast majority of other enzymes, incorporating only a few molecules of carbon dioxide into ribulose-1,5-bisphosphate per minute, but does so at atmospheric pressure and in mild, biological conditions. The resulting product is further reduced and eventually used in the synthesis of glucose, which in turn is a precursor to more complex carbohydrates, such as cellulose and starch. The process consumes energy in the form of ATP and NADPH.

Artificial CO_2 reduction for fuel production aims mostly at producing reduced carbon compounds from atmospheric CO_2. Some transition metal polyphosphine complexes have been developed for this end; however, they usually require previous concentration of CO_2 before use, and carriers (molecules that would fixate CO_2) that are both stable in aerobic conditions and able to concentrate CO_2 at atmospheric concentrations haven't been yet developed. The simplest product from CO_2 reduction is carbon monoxide (CO), but for fuel development, further reduction is needed, and a key step also needing further development is the transfer of hydride anions to CO.

Other Materials and Components

Charge separation is a major property of dyad and triad assemblies. Some nanomaterials employed are fullerenes (such as carbon nanotubes), a strategy that explores the pi-bonding properties of these materials. Diverse modifications (covalent and non-covalent) of carbon nanotubes have been attempted to increase the efficiency of charge separation, including the addition of ferrocene and pyrrole-like molecules such as porphyrins and phthalocyanines.

Since photodamage is usually a consequence in many of the tested systems after a period of exposure to light, bio-inspired photoprotectants have been tested, such as carotenoids (which are used in photosynthesis as natural protectants).

Light-driven Methodologies under Development

Photoelectrochemical Cells

Photoelectrochemical cells are a heterogeneous system that use light to produce either electricity or hydrogen. The vast majority of photoelectrochemical cells use semiconductors as catalysts. There have been attempts to use synthetic manganese complex-impregnated Nafion as a working electrode, but it has been since shown that the catalytically active species is actually the broken-down complex.

A promising, emerging type of solar cell is the dye-sensitized solar cell. This type of cell still depends on a semiconductor (such as TiO_2) for current conduction on one electrode, but with a coating of an organic or inorganic dye that acts as a photosensitizer; the counter electrode is a platinum catalyst for H_2 production. These cells have a self-repair mechanism and solar-to-electricity conversion efficiencies rivaling those of solid-state semiconductor ones.

Photocatalytic Water Splitting in Homogeneous Systems

Direct water oxidation by photocatalysts is a more efficient usage of solar energy than photoelectrochemical water splitting because it avoids an intermediate thermal or electrical energy conversion step.

Bio-inspired manganese clusters have been shown to possess water oxidation activity when adsorbed on clays together with ruthenium photosensitizers, although with low turnover numbers.

As mentioned above, some ruthenium complexes are able to oxidize water under solar light irradiation. Although their photostability is still an issue, many can be reactivated by a simple adjustment of the conditions in which they work. Improvement of catalyst stability has been tried resorting to polyoxometalates, in particular ruthenium-based ones. Another way to achieve improved stability may be the use of robust clathrochelate ligands that stabilize high oxidation states of metal in catalytic intremediates.

Whereas a fully functional artificial system is usually intended when constructing a water splitting device, some mixed methods have been tried. One of these involve the use of a gold electrode to which photosystem II is linked; an electric current is detected upon illumination.

Hydrogen-producing Artificial Systems

A H-cluster FeFe hydrogenase model compound covalently linked to a ruthenium photosensitizer. The ruthenium complex absorbs light and transduces its energy to the iron compound, which can then reduce protons to H_2.

The simplest photocatalytic hydrogen production unit consists of a hydrogen-evolving catalyst linked to a photosensitizer. In this dyad assembly, a so-called sacrificial donor for the photosensitizer is needed, that is, one that is externally supplied and replenished; the photosensitizer donates the necessary reducing equivalents to the hydrogen-evolving catalyst, which uses protons from a solution where it is immersed or dissolved in. Cobalt compounds such as cobaloximes are some of the best hydrogen catalysts, having been coupled to both metal-containing and metal-free photosensitizers. The first H-cluster models linked to photosensitizers (mostly ruthenium photosensitizers, but also porphyrin-derived ones) were prepared during the early 2000s. Both types of assembly are under development to improve their stability and increase their turnover numbers, both necessary for constructing a sturdy, long-lived solar fuel cell.

As with water oxidation catalysis, not only fully artificial systems have been idealized: hydrogenase enzymes themselves have been engineered for photoproduction of hydrogen, by coupling the enzyme to an artificial photosensitizer, such as [Ru(bipy)$_3$]$^{2+}$ or even photosystem I.

NADP$^+$/NADPH Coenzyme-inspired Catalyst

In natural photosynthesis, the NADP+ coenzyme is reducible to NADPH through binding of a proton and two electrons. This reduced form can then deliver the proton and electrons, potentially as a hydride, to reactions that culminate in the production of carbohydrates (the Calvin cycle). The coenzyme is recyclable in a natural photosynthetic cycle, but this process is yet to be artificially replicated.

A current goal is to obtain an NADPH-inspired catalyst capable of recreating the natural cyclic process. Utilizing light, hydride donors would be regenerated and produced where the molecules are continuously used in a closed cycle. Brookhaven chemists are now using a ruthenium-based complex to serve as the acting model. The complex is proven to perform correspondingly with NADP+/NADPH, behaving as the foundation for the proton and two electrons needed to convert acetone to isopropanol.

Currently, Brookhaven researchers are aiming to find ways for light to generate the hydride donors. The general idea is to use this process to produce fuels from carbon dioxide.

Photobiological Production of Fuels

Some photoautotrophic microorganisms can, under certain conditions, produce hydrogen. Nitrogen-fixing microorganisms, such as filamentous cyanobacteria, possess the enzyme nitrogenase, responsible for conversion of atmospheric N_2 into ammonia; molecular hydrogen is a byproduct of this reaction, and is many times not released by the microorganism, but rather taken up by a hydrogen-oxidizing (uptake) hydrogenase. One way of forcing these organisms to produce hydrogen is then to annihilate uptake hydrogenase activity. This has been done on a strain of *Nostoc punctiforme*: one of the structural genes of the NiFe uptake hydrogenase was inactivated by insertional mutagenesis, and the mutant strain showed hydrogen evolution under illumination.

Many of these photoautotrophs also have bidirectional hydrogenases, which can produce hydrogen under certain conditions. However, other energy-demanding metabolic pathways can compete with the necessary electrons for proton reduction, decreasing the efficiency of the overall process; also, these hydrogenases are very sensitive to oxygen.

Several carbon-based biofuels have also been produced using cyanobacteria, such as 1-butanol.

Synthetic biology techniques are predicted to be useful for this topic. Microbiological and enzymatic engineering have the potential of improving enzyme efficiency and robustness, as well as constructing new biofuel-producing metabolic pathways in photoautotrophs that previously lack them, or improving on the existing ones. Another topic being developed is the optimization of photobioreactors for commercial application.

Employed Research Techniques

Research in artificial photosynthesis is necessarily a multidisciplinary topic, requiring a multitude of different expertise. Some techniques employed in making and investigating catalysts and solar cells include:

- Organic and inorganic chemical synthesis.

- Electrochemistry methods, such as photoelectrochemistry, cyclic voltamme-try, electrochemical impedance spectroscopy Dielectric spectroscopy, and bulk electrolysis.

- Spectroscopic methods:

    ○ Fast techniques, such as time-resolved spectroscopy and ultrafast laser spectroscopy;

    ○ Magnetic resonance spectroscopies, such as nuclear magnetic resonance, electron paramagnetic resonance;

    ○ X-ray spectroscopy methods, including x-ray absorption such as XANES and EXAFS, but also x-ray emission.

- Crystallography.

- Molecular biology, microbiology and synthetic biology methodologies.

Advantages, Disadvantages and Efficiency

Advantages of solar fuel production through artificial photosynthesis include:

- The solar energy can be immediately converted and stored. In photovoltaic cells, sunlight is converted into electricity and then converted again into chemical energy for storage, with some necessary loss of energy associated with the second conversion.

- The byproducts of these reactions are environmentally friendly. Artificially photosynthesized fuel would be a carbon-neutral source of energy, which could be used for transportation or homes.

Disadvantages include:

- Materials used for artificial photosynthesis often corrode in water, so they may be less stable than photovoltaics over long periods of time. Most hydrogen catalysts are very sensitive to oxygen, being inactivated or degraded in its presence; also, photodamage may occur over time.

- The cost is not (yet) advantageous enough to compete with fossil fuels as a commercially viable source of energy.

A concern usually addressed in catalyst design is efficiency, in particular how much of the incident light can be used in a system in practice. This is comparable with photosynthetic efficiency, where light-to-chemical-energy conversion is measured. Photosynthetic organisms are able to collect about 50% of incident solar radiation, however the theoretical limit of photosynthetic efficiency is 4.6 and 6.0% for C_3 and C_4 plants respectively. In reality, the efficiency of photosynthesis is much lower and is usually below 1%, with some exceptions such as sugarcane in tropical climate. In contrast, the highest reported efficiency for artificial photosynthesis lab prototypes is 22.4%. However, plants are efficient in using CO_2 at atmospheric concentrations, something that artificial catalysts still cannot perform.

LIGHT-HARVESTING COMPLEX

A light-harvesting complex has a complex of subunit proteins that may be part of a larger supercomplex of a photosystem, the functional unit in photosynthesis. It is used by plants and photosynthetic bacteria to collect more of the incoming light than would be captured by the photosynthetic reaction center alone. Light-harvesting complexes are found in a wide variety among the different photosynthetic species. The complexes consist of proteins and photosynthetic pigments and surround a photosynthetic reaction center to focus energy, attained from photons absorbed by the pigment, toward the reaction center using Förster resonance energy transfer.

Function

Absorption of a photon by a molecule takes place leading to electronic excitation when the energy of the captured photon matches that of an electronic transition. The fate of such excitation can be a return to the ground state or another electronic state of the same molecule. When the excited molecule has a nearby neighbour molecule, the excitation energy may also be transferred, through electromagnetic interactions, from one molecule to another. This process is called resonance energy transfer, and the rate depends strongly on the distance between the energy donor and energy acceptor molecules. Light-harvesting complexes have their pigments specifically positioned to optimize these rates.

In Purple Bacteria

Purple bacteria use bacteriochlorophyll and caretonoids to gather light energy. These proteins are arranged in a ring-like fashion creating a cylinder that spans the membrane.

In Green Bacteria

Green sulphur bacteria and some Chloroflexia use ellipsoidal complexes known as the chlorosome to capture light. Their form of bacteriochlorophyll is green.

In Cyanobacteria and Plants

Chlorophylls and carotenoids are important in light-harvesting complexes present in plants. Chlorophyll b is almost identical to chlorophyll a except it has a formyl group in place of a methyl group. This small difference makes chlorophyll b absorb light with wavelengths between 400 and 500 nm more efficiently. Carotenoids are long linear organic molecules that have alternating single and double bonds along their length. Such molecules are called polyenes. Two examples of carotenoids are lycopene and β-carotene. These molecules also absorb light most efficiently in the 400 – 500 nm range. Due to their absorption region, carotenoids appear red and yellow and provide most of the red and yellow colours present in fruits and flowers.

The carotenoid molecules also serve a safeguarding function. Carotenoid molecules suppress damaging photochemical reactions, in particular those including oxygen, which exposure to sunlight can cause. Plants that lack carotenoid molecules quickly die upon exposure to oxygen and light.

Phycobilisome

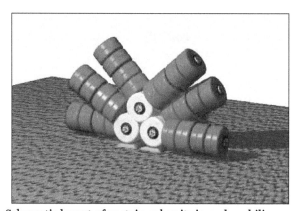

Schematic layout of protein subunits in a phycobilisome.

Little light reaches algae that reside at a depth of one meter or more in seawater, as light is absorbed by seawater. A phycobilisome is a light-harvesting protein complex present in cyanobacteria, glaucocystophyta, and red algae and is structured like a real antenna. The pigments, such as phycocyanobilin and phycoerythrobilin, are the chromophores that bind through a covalent thioether bond to their apoproteins at cysteins residues. The apoprotein with its chromophore is called phycocyanin, phycoerythrin, and allophycocyanin, respectively. They often occur as hexamers of α and β subunits $(\alpha_3\beta_3)_2$. They enhance the amount and spectral window of light absorption and fill the "green gap", which occur in higher Plants.

The geometrical arrangement of a phycobilisome is very elegant and results in 95% efficiency of energy transfer. There is a central core of allophycocyanin, which sits above a photosynthetic reaction center. There are phycocyanin and phycoerythrin subunits that radiate out from this center like thin tubes. This increases the surface area of the absorbing section and helps focus and concentrate light energy down into the reaction center to a Chlorophyll. The energy transfer from excited electrons absorbed by pigments in the phycoerythrin subunits at the periphery of these antennas appears at the reaction center in less than 100 ps.

NON-PHOTOCHEMICAL QUENCHING

Non-photochemical quenching (NPQ) is a mechanism employed by plants and algae to protect themselves from the adverse effects of high light intensity. It involves the quenching of singlet excited state chlorophylls (Chl) via enhanced internal conversion to the ground state (non-radiative decay), thus harmlessly dissipating excess excitation energy as heat through molecular vibrations. NPQ occurs in almost all photosynthetic eukaryotes (algae and plants), and helps to regulate and protect photosynthesis in environments where light energy absorption exceeds the capacity for light utilization in photosynthesis.

Process

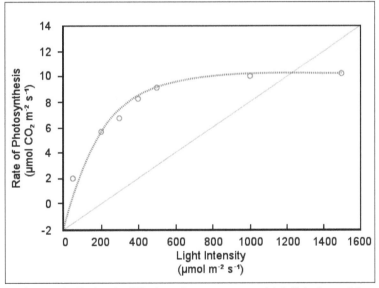

Carbon assimilation (red line) tends to saturate at high light intensities, while light absorption (blue line) increases linearly.

When a molecule of chlorophyll absorbs light it is promoted from its ground state to its first singlet excited state. The excited state then has three main fates. Either the energy is; 1. passed to another chlorophyll molecule by Förster resonance energy transfer (in this way excitation is gradually passed to the photochemical reaction centers (photosystem I and photosystem II) where energy is used in photosynthesis (called photochemical quenching); or 2. the excited state can return to the ground state by emitting the energy as heat (called non-photochemical quenching); or 3. the excited state can return to the ground state by emitting a photon (fluorescence).

In higher plants, the absorption of light continues to increase as light intensity increases, while the capacity for photosynthesis tends to saturate. Therefore, there is the potential for the absorption of excess light energy by photosynthetic light harvesting systems. This excess excitation energy leads to an increase in the lifetime of singlet excited chlorophyll, increasing the chances of the formation of long-lived chlorophyll triplet states by inter-system crossing. Triplet chlorophyll is a potent photosensitiser of molecular oxygen forming singlet oxygen which can cause oxidative damage to the pigments, lipids and proteins of the photosynthetic thylakoid membrane. To counter this problem, one photoprotective mechanism is so-called non-photochemical quenching (NPQ), which relies upon the conversion and dissipation of the excess excitation energy into heat. NPQ involves conformational changes within the light harvesting proteins of photosystem (PS) II that bring about a change in pigment interactions causing the formation of energy traps. The conformational changes are stimulated by a combination of transmembrane proton gradient, the PsbS subunit of photosystem II and the enzymatic conversion of the carotenoid violaxanthin to zeaxanthin (the xanthophyll cycle).

Violaxanthin is a carotenoid downstream from chlorophyll a and b within the antenna of PS II and nearest to the special chlorophyll a located in the reaction center of the antenna. As light intensity increases, acidification of the thylakoid lumen takes place. This acidification leads to the protonation of the PsbS subunit of PS II. Along with the activation of enzyme violaxanthin de-epoxidase which eliminates an epoxide and forms an alkene on a six-member ring of violaxanthin giving another carotenoid known as antheraxanthin. Violaxanthin contains two epoxides each bonded to a six-member ring and when both are eliminated by de-epoxidase the carotenoid zeaxanthin is formed. Only violaxanthin is able to transport a photon to the special chlorophyll a. Antheraxanthin and zeaxanthin dissipate the energy from the photon as heat preserving the integrity of photosystem II. This dissipation of energy as heat is one form of non-photochemical quenching.

Measurement of NPQ

Non-photochemical quenching is measured by the quenching of chlorophyll fluorescence and is distinguished from photochemical quenching by applying a bright light pulse to transiently saturate photochemical quenching thus removing its contribution from the observed quenching. Non-photochemical quenching is not affected if the pulse of light is short. During the pulse, the fluorescence reaches the level reached in the absence of any photochemical quenching, known as maximum fluorescence.

Chlorophyll fluorescence can easily be measured with a chlorophyll fluorometer. Some fluorometers can calculate NPQ and photochemical quenching coefficients (including qP, qN, qE and NPQ), as well as light and dark adaptation parameters (including Fo, Fm, and Fv/Fm).

PHOTOINHIBITION

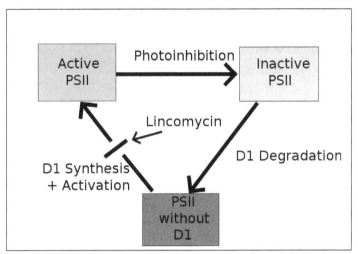

Photoinhibition of Photosystem II (PSII) leads to loss of PSII electron transfer activity.
PSII is continuously repaired via degradation and synthesis of the D1 protein.
Lincomycin can be used to block protein synthesis.

Photoinhibition is light-induced reduction in the photosynthetic capacity of a plant, alga, or cyanobacterium. Photosystem II (PSII) is more sensitive to light than the rest of the photosynthetic machinery, and most researchers define the term as light-induced damage to PSII. In living organisms, photoinhibited PSII centres are continuously repaired via degradation and synthesis of the D1 protein of the photosynthetic reaction center of PSII. Photoinhibition is also used in a wider sense, as dynamic photoinhibition, to describe all reactions that decrease the efficiency of photosynthesis when plants are exposed to light.

Photoinhibition occurs in all organisms capable of oxygenic photosynthesis, from vascular plants to cyanobacteria. In both plants and cyanobacteria, blue light causes photoinhibition more efficiently than other wavelengths of visible light, and all wavelengths of ultraviolet light are more efficient than wavelengths of visible light. Photoinhibition is a series of reactions that inhibit different activities of PSII, but there is no consensus on what these steps are. The activity of the oxygen-evolving complex of PSII is often found to be lost before the rest of the reaction centre loses activity. However, inhibition of PSII membranes under anaerobic conditions leads primarily to inhibition of electron transfer on the acceptor side of PSII. Ultraviolet light causes inhibition of the oxygen-evolving complex before the rest of PSII becomes inhibited. Photosystem I (PSI) is less susceptible to light-induced damage than PSII, but slow inhibition of this photosystem has been observed. Photoinhibition of PSI occurs in chilling-sensitive plants and the reaction depends on electron flow from PSII to PSI.

Photosystem II is damaged by light irrespective of light intensity. The quantum yield of the damaging reaction in typical leaves of higher plants exposed to visible light, as well as in isolated thylakoid membrane preparations, is in the range of 10^{-8} to 10^{-7} and independent of the intensity of light. This means that one PSII complex is damaged for every 10-100 million photons that are intercepted. Therefore, photoinhibition occurs at all light intensities and the rate constant of photoinhibition is directly proportional to light intensity. Some measurements suggest that dim light causes damage more efficiently than strong light.

Molecular Mechanisms

The mechanisms of photoinhibition are under debate, several mechanisms have been suggested. Reactive oxygen species, especially singlet oxygen, have a role in the acceptor-side, singlet oxygen and low-light mechanisms. In the manganese mechanism and the donor side mechanism, reactive oxygen species do not play a direct role. Photoinhibited PSII produces singlet oxygen, and reactive oxygen species inhibit the repair cycle of PSII by inhibiting protein synthesis in the chloroplast.

Acceptor-side Photoinhibition

Strong light causes the reduction of the plastoquinone pool, which leads to protonation and double reduction (and double protonation) of the Q_A electron acceptor of Photosystem II. The protonated and double-reduced forms of Q_A do not function in electron transport. Furthermore, charge recombination reactions in inhibited Photosystem II are expected to lead to the triplet state of the primary donor (P_{680}) more probably than same reactions in active PSII. Triplet P_{680} may react with oxygen to produce harmful singlet oxygen.

Donor-side Photoinhibition

If the oxygen-evolving complex is chemically inactivated, then the remaining electron transfer activity of PSII becomes very sensitive to light. It has been suggested that even in a healthy leaf, the oxygen-evolving complex does not always function in all PSII centers, and those ones are prone to rapid irreversible photoinhibition.

Manganese Mechanism

A photon absorbed by the manganese ions of the oxygen-evolving complex triggers inactivation of the oxygen-evolving complex. Further inhibition of the remaining electron transport reactions occurs like in the donor-side mechanism. The mechanism is supported by the action spectrum of photoinhibition.

Singlet Oxygen Mechanisms

Inhibition of PSII is caused by singlet oxygen produced either by weakly coupled chlorophyll molecules or by cytochromes or iron-sulfur centers.

Low-light Mechanism

Charge recombination reactions of PSII cause the production of triplet P_{680} and, as a consequence, singlet oxygen. Charge recombination is more probable under dim light than under higher light intensities.

Kinetics and Action Spectrum

Photoinhibition follows simple first-order kinetics if measured from a lincomycin-treated leaf, cyanobacterial or algal cells, or isolated thylakoid membranes in which concurrent repair does not disturb the kinetics. Data from the group of W. S. Chow indicate that in leaves of pepper (*Capsicum annuum*), the first-order pattern is replaced by a pseudo-equilibrium even if the repair reaction is blocked. The deviation has been explained by assuming that photoinhibited PSII centers protect the remaining active ones. Both visible and ultraviolet light cause photoinhibition, ultraviolet wavelengths being much more damaging. Some researchers consider ultraviolet and visible light induced photoinhibition as a two different reactions, while others stress the similarities between the inhibition reactions occurring under different wavelength ranges.

PSII Repair Cycle

Photoinhibition occurs continuously when plants or cyanobacteria are exposed to light, and the photosynthesizing organism must, therefore, continuously repair the damage. The PSII repair cycle, occurring in chloroplasts and in cyanobacteria, consists of degradation and synthesis of the D1 protein of the PSII reaction centre, followed by activation of the reaction center. Due to the rapid repair, most PSII reaction centers are not photoinhibited even if a plant is grown in strong light. However, environmental stresses, for example, extreme temperatures, salinity, and drought, limit the supply of carbon dioxide for use in carbon fixation, which decreases the rate of repair of PSII.

In photoinhibition studies, repair is often stopped by applying an antibiotic (lincomycin or chloramphenicol) to plants or cyanobacteria, which blocks protein synthesis in the chloroplast. Protein synthesis occurs only in an intact sample, so lincomycin is not needed when photoinhibition is measured from isolated membranes. The repair cycle of PSII recirculates other subunits of PSII (except for the D1 protein) from the inhibited unit to the repaired one.

Protective Mechanisms

Plants have mechanisms that protect against adverse effects of strong light. The most studied biochemical protective mechanism is non-photochemical quenching of excitation energy. Visible-light-induced photoinhibition is ~25% faster in an *Arabidopsis thaliana* mutant lacking non-photochemical quenching than in the wild type. It is also apparent that turning or folding of leaves, as occurs, e.g., in *Oxalis* species in response to exposure to high light, protects against photoinhibition.

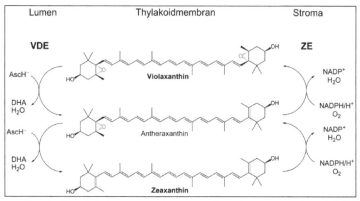

The xanthophyll cycle is important in protecting plants from photoinhibition.

The PsBs Protein

Because there are a limited number of photosystems in the electron transport chain, organisms that are photosynthetic must find a way to combat excess light and prevent photo-oxidative stress, and likewise, photoinhibition, at all costs. In an effort to avoid damage to the D1 subunit of PSII and subsequent formation of ROS, the plant cell employs accessory proteins to carry the excess excitation energy from incoming sunlight; namely, the PsBs protein. Elicited by a relatively low luminal pH, plants have developed a rapid response to excess energy by which it is given off as heat and damage is reduced.

The studies of Tibiletti *et al.* found that PsBs is the main protein involved in sensing the changes in the pH and can therefore rapidly accumulate in the presence of high light. This was determined by performing SDS-PAGE and immunoblot assays, locating PsBs itself in the green alga, *Chlamydomonas reinhardtii*. Their data concluded that the PsBs protein belongs to a multigene family termed LhcSR proteins, including the proteins that catalyze the conversion of violaxanthin to zeaxanthin, as previously mentioned. PsBs is involved in the changing the orientation of the photosystems at times of high light to prompt the arrangement of a quenching site in the light harvesting complex.

Additionally, studies conducted by Glowacka *et al.* show that a higher concentration of PsBs is directly correlated to inhibiting stomatal aperture. But it does this without affecting CO_2 intake and it increases water use efficiency of the plant. This was determined by controlling the expression of PsBs in *Nicotinana tabacum* by imposing a series of genetic modifications to the plant in order to test for PsBs levels and activity including: DNA transformation and transcription followed by protein expression. Research shows that stomatal conductance is heavily dependent on the presence of the PsBs protein. Thus, when PsBs was overexpressed in a plant, water uptake efficiency was seen to significantly improve, resulting in new methods for prompting higher, more productive crop yields.

These recent discoveries tie together two of the largest mechanisms in phytobiology; these are the influences that the light reactions have upon stomatal aperture via the Calvin Benson Cycle. To elaborate, the Calvin-Benson Cycle, occurring in the stroma of the chloroplast obtains its CO_2 from the atmosphere which enters upon stomatal opening. The energy to drive the Calvin-Benson cycle is a product of the light reactions. Thus, the relationship has been discovered as such: when PsBs is silenced, as expected, the excitation pressure at PSII is increased. This in turn results in an activation of the redox state of Quinone A and there is no change in the concentration of carbon dioxide in the intracellular airspaces of the leaf; ultimately increasing stomatal conductance. The inverse relationship also holds true: when PsBs is over expressed, there is a decreased excitation pressure at PSII. Thus, the redox state of Quinone A is no longer active and there is, again, no change in the concentration of carbon dioxide in the intracellular airspaces of the leaf. All these factors work to have a net decrease of stomatal conductance.

Measurement

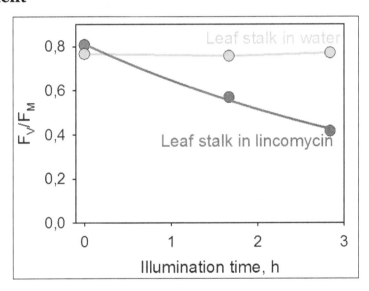

Effect of illumination on the ratio of variable to maximum fluorescence (F_V/F_M) of ground-ivy (*Glechoma hederacea*) leaves. Photon flux density was 1000 µmol m^{-2}s^{-1}, corresponding to half of full sunlight. Photoinhibition damages PSII at the same rate whether the leaf stalk is in water or lincomycin, but, in the "leaf stalk in water" sample, repair is so rapid that no net decrease in (F_V/F_M) occurs.

Photoinhibition can be measured from isolated thylakoid membranes or their subfractions, or from intact cyanobacterial cells by measuring the light-saturated rate of oxygen evolution in the presence of an artificial electron acceptor (quinones and dichlorophenol-indophenol have been used).

The degree of photoinhibition in intact leaves can be measured using a fluorimeter to measure the ratio of variable to maximum value of chlorophyll a fluorescence (F_V/F_M). This ratio can be used as a proxy of photoinhibition because more energy is emitted as fluorescence from Chlorophyll a when many excited electrons from PSII are not captured by the acceptor and decay back to their ground state.

When measuring F_V/F_M, the leaf must be incubated in the dark for at least 10 minutes, preferably longer, before the measurement, in order to let non-photochemical quenching relax.

Flashing Light

Photoinhibition can also be induced with short flashes of light using either a pulsed laser or a xenon flash lamp. When very short flashes are used, the photoinhibitory efficiency of the flashes depends on the time difference between the flashes. This dependence has been interpreted to indicate that the flashes cause photoinhibition by inducing recombination reactions in PSII, with subsequent production of singlet oxygen. The interpretation has been criticized by noting that the photoinhibitory efficiency of xenon flashes depends on the energy of the flashes even if such strong flashes are used that they would saturate the formation of the substrate of the recombination reactions.

Dynamic Photoinhibition

Some researchers prefer to define the term "photoinhibition" so that it contains all reactions that lower the quantum yield of photosynthesis when a plant is exposed to light. In this case, the term "dynamic photoinhibition" comprises phenomena that reversibly down-regulate photosynthesis in the light and the term "photodamage" or "irreversible photoinhibition" covers the concept of photoinhibition used by other researchers. The main mechanism of dynamic photoinhibition is non-photochemical quenching of excitation energy absorbed by PSII. Dynamic photoinhibition is acclimation to strong light rather than light-induced damage, and therefore "dynamic photoinhibition" may actually protect the plant against "photoinhibition".

THYLAKOID

A thylakoid is a sheet-like membrane-bound structure that is the site of the light-dependent photosynthesis reactions in chloroplasts and cyanobacteria. It is the site that contains the chlorophyll used to absorb light and use it for biochemical reactions. Thylakoids may also be called lamellae, although this term may be used to refer to the portion of a thylakoid that connects grana.

Thylakoid Structure

In chloroplasts, thylakoids are embedded in the stroma (interior portion of a chloroplast). The stroma contains ribosomes, enzymes, and chloroplast DNA. The thylakoid consists of the thylakoid membrane and the enclosed region called the thylakoid lumen. A stack of thylakoids forms a group of coin-like structures called a granum. A chloroplast contains several of these structures, collectively known as grana.

Higher plants have specially organized thylakoids in which each chloroplast has 10–100 grana that are connected to each other by stroma thylakoids. The stroma thylakoids may be thought of as tunnels that connect the grana. The grana thylakoids and stroma thylakoids contain different proteins.

Role of the Thylakoid in Photosynthesis

Reactions performed in the thylakoid include water photolysis, the electron transport chain, and ATP synthesis.

Photosynthetic pigments (e.g., chlorophyll) are embedded into the thylakoid membrane, making it the site of the light-dependent reactions in photosynthesis. The stacked coil shape of the grana gives the chloroplast a high surface area to volume ratio, aiding the efficiency of photosynthesis.

The thylakoid lumen is used for photophosphorylation during photosynthesis. The light-dependent reactions in the membrane pump protons into the lumen, lowering its pH to 4. In contrast, the pH of the stroma is 8.

Water Photolysis

The first step is water photolysis, which occurs on the lumen site of the thylakoid membrane. Energy from light is used to reduce or split water. This reaction produces electrons that are needed for the electron transport chains, protons that are pumped into the lumen to produce a proton gradient, and oxygen. Although oxygen is needed for cellular respiration, the gas produced by this reaction is returned to the atmosphere.

Electron Transport Chain

The electrons from photolysis go to the photosystems of the electron transport chains. The photosystems contain an antenna complex that uses chlorophyll and related pigments to collect light at various wavelengths. Photosystem I uses light to reduce $NADP^+$ to produce NADPH and H^+. Photosystem II uses light to oxidize water to produce molecular oxygen (O_2), electrons (e^-), and protons (H^+). The electrons reduce $NADP^+$ to NADPH in both systems.

ATP Synthesis

ATP is produced from both Photosystem I and Photosystem II. Thylakoids synthesize

ATP using an ATP synthase enzyme that is similar to mitochondrial ATPase. The enzyme is integrated into the thylakoid membrane. The CF1-portion of the synthase molecule extended into the stroma, where ATP supports the light-independent photosynthesis reactions.

The lumen of the thylakoid contains proteins used for protein processing, photosynthesis, metabolism, redox reactions, and defense. The protein plastocyanin is an electron transport protein that transports electrons from the cytochrome proteins to Photosystem I. Cytochrome b_6f complex is a portion of the electron transport chain that couples proton pumping into the thylakoid lumen with electron transfer. The cytochrome complex is located between Photosystem I and Photosystem II.

Thylakoids in Algae and Cyanobacteria

While thylakoids in plant cells form stacks of grana in plants, they may be unstacked in some types of algae.

While algae and plants are eukaryotes, cyanobacteria are photosynthetic prokaryotes. They do not contain chloroplasts. Instead, the entire cell acts as a sort of thylakoid. The cyanobacterium has an outer cell wall, cell membrane, and thylakoid membrane. Inside this membrane is the bacterial DNA, cytoplasm, and carboxysomes. The thylakoid membrane has functional electron transfer chains that support photosynthesis and cellular respiration. Cyanobacteria thylakoid membranes don't form grana and stroma. Instead, the membrane forms parallel sheets near the cytoplasmic membrane, with enough space between each sheet for phycobilisomes, the light harvesting structures.

References

- Milo, Ron; Philips, Rob. "Cell Biology by the Numbers: How large are chloroplasts?". Book.bionumbers.org. Retrieved 7 February 2017

- Plastids, biology: byjus.com, Retrieved 4 February, 2019

- Oostergetel GT, van Amerongen H, Boekema EJ (June 2010). "The chlorosome: a prototype for efficient light harvesting in photosynthesis". Photosynthesis Research. 104 (2–3): 245–55. Doi:10.1007/s11120-010-9533-0. PMC 2882566. PMID 20130996

- Krause GH & Jahns P (2004) "Non-photochemical energy dissipation determined by chlorophyll fluorescence quenching: Characterization and function" in Papageorgiou GC & Govindjee (eds.) "Chlorophyll Fluorescence: A Signature of Photosynthesis". Pp. 463–495. Springer, The Netherlands. ISBN 978-1-4020-3217-2

- Thylakoid-definition-and-function-4125710: thoughtco.com, Retrieved 5 March, 2019

- Timmer J (7 December 2017). "We may now be able to engineer the most important lousy enzyme on the planet". Ars Technica. Retrieved 5 January 2019

3

Photosynthetic Pigments and Enzymes

The pigments which are used to capture the light energy which is essential for photosynthesis are known as photosynthetic pigments. Some of the major photosynthetic pigments and enzymes are chlorophyll, bacteriochlorophyll, allophycocyanin, pheophytin and phycobilin. This chapter has been carefully written to provide an easy understanding of the varied facets of these photosynthetic pigments and enzymes.

CHLOROPHYLL

Chlorophyll gives leaves their green color.

Chlorophyll is a green photosynthetic pigment found in plants, algae, and cyanobacteria. Chlorophyll is an essential component of photosynthesis, which helps plants get energy from light.

Chlorophyll absorbs most strongly in the blue and to a lesser extent red portions of the electromagnetic spectrum. The green portions of the electromagnetic spectrum, including wavelengths between five hundred and approximately six hundred nanometers, are not absorbed well, but are reflected by the plants, hence the green color of chloro-phyll-containing tissues like plant leaves.

Chlorophyll, via its central role in photosynthesis, reflects harmony on both the sub-cellular and macro levels. On the sub-cellular level, the conversion of light energy via chlorphyll into useable chemical energy requies the complex coordination of several parts. On the higher level, the harmony between green plants, animals, and the environment is seen in the fact that plants use chlorophyll and water to convert carbon dioxide into glucose and free oxygen, while animals correspondingly use the oxygen and trapped energy stored in plant biomass and return carbon dioxide.

Chlorophyll and Photosynthesis

In plant photosynthesis, incoming light is absorbed by chlorophyll and other accessory pigments in the antenna complexes of photosystem I and photosystem II. The antenna pigments are predominantly chlorophyll a, chlorophyll b, absorbing violet-blue and red light, respectively, and carotenoids. Their absorption spectra are non-overlapping, which serves to broaden the specific bandwidths of light these individual compounds absorb during the process of photosynthesis. The carotenoids also play a role as antioxidants, and serve to reduce photo-oxidative damage to chlorophyll molecules.

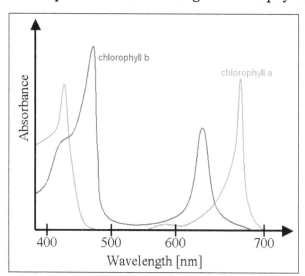

Absorbance spectra of free chlorophyll *a* (green) and *b* (red) in a solvent.
The spectra of chlorophyll molecules are slightly modified *in vivo*
depending on specific pigment-protein interactions.

Each antenna complex has between 250 and 400 pigment molecules, and the energy they absorb is shuttled by resonance energy transfer to a specialized chlorophyll *a* at the reaction center of each photosystem. When either of the two chorophyll *a* molecules at the reaction center absorb energy, an electron is excited and transferred to an electron-acceptor molecule, leaving an electron hole in the donor chlorophyll. In a poorly-understood reaction, electrons from water molecules participate in an oxidation reaction, where the hole from the donor chlorophyll is filled (recombined with another electron), and diatomic oxygen (O_2) is produced. Resulting chemical energy originating from the initial excited electron is eventually captured in the form of ATP and NADPH, and is then ultimately used to convert carbon dioxide (CO_2) to carbohydrates. This CO_2 fixation process results in the conversion (or an integrated external quantum efficiency) of 3 to 6 percent of the total incident solar radiation, with a theoretical maximum efficiency of 11 percent.

In Photosystem II, the electron which reduces P680+ ultimately comes from the oxidation of water into O_2 and H+ through several intermediates. This reaction is how photosynthetic organisms like plants produce O_2 gas, and is the source for practically all the O_2 in Earth's atmosphere.

Special Pair

The photosystem reaction centers consist of a "special pair" of chlorophyll *a* molecules that are characterized by their specific absorption maximum. The special pair in photosystem I are designated P700, and those from photosystem II are designated P680. The P is short for pigment, and the number is the specific absorption peak in nanometers for the chlorophyll molecules in each reaction center.

Chlorophyll *a* is common to all eukaryotic photosynthetic organisms, and, due to its central role in the reaction center, is essential for photosynthesis. The accessory pigments such as chlorophyll *b* and carotenoids are not essential. Some algae, such as brown algae and diatoms, use chlorophyll **c** as a substitute for chlorophyll *b*. Historically, red algae have been assumed to have chlorophyll d, although it could not be isolated from all species and even different collections of the same species. This puzzle has recently been resolved, since the chlorophyll *d* is actually from an epiphytic cyanobacterium (*Acaryochloris marina*) that lives on the red algae. These cyanobacteria have a ratio of chlorophyll *d*: chlorophyll *a* of approximately 30 to 1, and represent a rare example of a photosystem with chlorophyll *d* at the reaction center of the photosystem. All other known eukaryotes and cyanobacteria use chlorophyll *a*. There are likely to be many chlorophyll-d containing organisms awaiting discovery, for example a free living form was recently found in the Salton Sea (a salt lake in the United States).

Other chemical variations of chlorophyll are found in photosynthetic bacteria, other than cyanobacteria (sometimes called blue-green algae). Purple bacteria use bacteriochlorophyll, which absorbs infrared light between 800 to 1000 nanometers, and the green sulphur bacteria use chlorobium chlorophyll. All known bacteria with bacteriochlorophyll have a form of photosynthesis that does not involve evolution of oxygen and so are called anoxyphotobacteria. There is a very large number of different bacteriochlorophylls in different anoxyphotobacteria, including one species which contains zinc, rather than the usual magnesium as the coordinated metal.

Because the different chlorophyll and non-chlorophyll pigments associated with the photosystems all have different spectra, the total absorption spectrum is broadened and flattened such that a wider range of red, orange, yellow, and blue light can be absorbed by plants and algae. Most photosynthetic organisms do not have pigments which absorb green light well, thus most remaining light under leaf canopies in forests or under water with abundant plankton is green, a spectral effect called the "green window." Some organisms, such as cyanobacteria and red algae, contain accessory phycobilin pigments that can absorb green light relatively well and thus they can exploit the little remaining green light in these habitats.

Chemical Structure

	Chlorophyll a	Chlorophyll b	Chlorophyll c1	Chlorophyll c2	Chlorophyll d
Molecular formula	$C_{55}H_{72}O_5N_4Mg$	$C_{55}H_{70}O_6N_4Mg$	$C_{35}H_{30}O_5N_4Mg$	$C_{35}H_{28}O_5N_4Mg$	$C_{54}H_{70}O_6N_4Mg$
C3 group	$-CH=CH_2$	$-CH=CH_2$	$-CH=CH_2$	$-CH=CH_2$	$-CHO$
C7 group	$-CH_3$	$-CHO$	$-CH_3$	$-CH_3$	$-CH_3$
C8 group	$-CH_2CH_3$	$-CH_2CH_3$	$-CH_2CH_3$	$-CH=CH_2$	$-CH_2CH_3$
C17 group	$-CH_2CH_2COO$-Phytyl	$-CH_2CH_2COO$-Phytyl	$-CH=CH-COOH$	$-CH=CHCOOH$	$-CH_2CH_2COO$-Phytyl
C17-C18 bond	Single	Single	Double	Double	Single
Occurrence	Universal	Mostly plants	Various algae	Various algae	Cyanobacteria

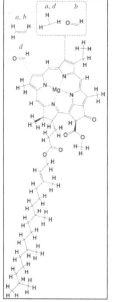

Common structure of chlorophyll a, band d.

Common structure of chlo-
rophyll c1, and c2.

Chlorophyll Biosynthesis

Plants make chlorophyll from the molecules glycine and succinyl-CoA. There is an intermediate molecule called protochlorophyllide, which is converted into chlorophyll. In angiosperms, this chemical reaction is light-dependent. These plants are pale if they are grown in darkness because they can't complete the reaction to produce chlorophyll. Algae and non-vascular plants don't require light to synthesize chlorophyll.

Protochlorophyllide forms toxic free radicals in plants, so chlorophyll biosynthesis is tightly regulated. If iron, magnesium, or iron are deficient, plants may be unable to synthesize enough chlorophyll, appearing pale or *chlorotic*. Chlorosis may also be caused by improper pH (acidity or alkalinity) or pathogens or insect attack.

Evidence for Chlorophyll

Chlorophyll can be shown to be vital for photosynthesis by destarching a leaf from a variegated plant and exposing it to light for several hours (variegated leaves have green areas that contain chlorophyll and white areas that have none). When tested with iodine solution, a color change revealing the presence of starch occurs only in regions of the leaf that were green and therefore contained chlorophyll. This shows that photosynthesis does not occur in areas where chlorophyll is absent, and constitutes evidence that the presence of chlorophyll is a requirement for photosynthesis.

Chlorosis

A corn plant with severe chlorosis (left) beside a normal plant (right).

In botany, chlorosis is a condition in which leaves produce insufficient chlorophyll. As chlorophyll is responsible for the green color of leaves, chlorotic leaves are pale, yellow, or yellow-white. The affected plant has little or no ability to manufacture carbohydrates through photosynthesis and may die unless the cause of its chlorophyll insufficiency is treated, although some chlorotic plants, such as the albino *Arabidopsis thaliana* mutant *ppi2*, are viable if supplied with exogenous sucrose.

In viticulture, the most common symptom of poor nutrition in grapevines is the yellowing of grape leaves caused by chlorosis and the subsequent loss of chlorophyll. This is often seen in vineyard soils that are high in limestone such as the Italian wine region of Barolo in the Piedmont, the Spanish wine region of Rioja and the French wine regions of Champagne and Burgundy. In these soils the grapevine often struggles to pull sufficient levels of iron which is a needed component in the production of chlorophyll.

Causes

A *Liquidambar* leaf with interveinal chlorosis.   Citron shrub with chlorosis.

Chlorosis is typically caused when leaves do not have enough nutrients to synthesise all the chlorophyll they need. It can be brought about by a combination of factors including:

- A specific mineral deficiency in the soil, such as iron, magnesium or zinc.

- Deficient nitrogen and/or proteins.

- A soil pH at which minerals become unavailable for absorption by the roots.

- Poor drainage (waterlogged roots).

- Damaged and/or compacted roots.

- Pesticides and particularly herbicides may cause chlorosis, both to target weeds and occasionally to the crop being treated.

- Exposure to sulphur dioxide.

- Ozone injury to sensitive plants.

- Presence of any number of bacterial pathogens, for instance *Pseudomonas syringae pv. tagetis* that causes complete chlorosis on Asteraceae.

- Fungal infection, e.g. Bakanae.

However, the exact conditions vary from plant type to plant type. For example, Azaleas grow best in acidic soil and rice is unharmed by waterlogged soil.

In Grape Vines

Like many other plants, grape vines are susceptible to chlorosis, and symptoms of iron deficiency tend to be common on soils rich in limestone. In the wake of The Great French Wine Blight, when European *Vitis vinifera* were affected by *Phylloxera*, chlorosis became a greater problem in viticulture. To deal with the Phylloxera blight, *V. vinifera* was grafted onto rootstock based on American species of the genus *Vitis*, such as *Vitis riparia, Vitis rupestris, Vitis berlandieri*. However, many of these were less adapted to the lime-rich soils that were common in France's vineyards, in particular many of those that produced wines of top quality. Many grafted vines in lime-rich vineyards therefore showed signs of iron deficiency, and in France this specific form of chlorosis was termed *chlorose calcaire*. The problem was largely overcome by the selection of lime-resistant American vines as basis for hybrid vines used for rootstock material. However, since such rootstocks may be less than optimal in other respects, it is necessary for the viticulturalist to balance the need for chlorosis resistance against other viticultural needs. This is illustrated by one of the most common lime-resistant rootstocks, 41 B, which is a hybrid between *V. vinifera* cultivar Chasselas and *V. berlandieri*, which generally has a sufficient, but not extremely high, Phylloxera resistance.

Treatments

Specific nutrient deficiencies (often aggravated by high soil pH) may be corrected by supplemental feedings of iron, in the form of a chelate or sulphate, magnesium or nitrogen compounds in various combinations.

BACTERIOCHLOROPHYLL

Bacteriochlorophyll is a form of chlorophyll found in photosynthetic bacteria, notably the purple and green bacteria. There are several types, designated *a* to *g*. For example, bacteriochlorophyll *a* and bacteriochlorophyll *b* are structurally similar to the chlorophyll *a* and chlorophyll *b* found in plants; the purple bacteria contain either of these two types of bacteriochlorophyll, depending on the species. Bacteriochlorophyll is located in specialized membrane systems (*chromatophores*); in purple bacteria these are in the form of sheets, tubes, or vesicles arising inside the cell from the plasma membrane, whereas in green bacteria they are cylindrical structures (*chlorosomes*) underlying the plasma membrane.

ALLOPHYCOCYANIN

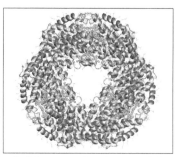

Allophycocyanin dodekamer, *Gloeobacter violaceus* (PDB: 2vjt).

Allophycocyanin is a protein from the light-harvesting phycobiliprotein family, along with phycocyanin, phycoerythrin and phycoerythrocyanin. It is an accessory pigment to chlorophyll. All phycobiliproteins are water-soluble and therefore cannot exist within the membrane like carotenoids, but aggregate forming clusters that adhere to the membrane called phycobilisomes. Allophycocyanin absorbs and emits red light (650 & 660 nm max, respectively), and is readily found in Cyanobacteria (also called blue-green algae), and red algae. Phycobilin pigments have fluorescent properties that are used in immunoassay kits. In flow cytometry, it is often abbreviated APC. To be effectively used in applications such as FACS, High-Throughput Screening (HTS) and microscopy, APC needs to be chemically cross-linked.

Structural Characteristics

Allophycocyanin can be isolated from various species of red or blue-green algae, each producing slightly different forms of the molecule. It is composed of two different subunits (α and β) in which each subunit has one phycocyanobilin (PCB) chromophore. The subunit structure for APC has been determined as $(\alpha\beta)_3$. The molecular weight of APC is 105,000 Daltons.

Spectral Characteristics

Absorption maximum	652 nm
Additional Absorption peak	625 nm
Emission maximum	657.5 nm
Stokes Shift	5.5 nm
Extinction Coefficient	700,000 $M^{-1}cm^{-1}$
Quantum Yield	0.68
Brightness	1.6 x 10^5 $M^{-1}cm^{-1}$

Cross-linked APC

As mentioned above, in order for APC to be useful in immunoassays it must first be chemically cross-linked to prevent it from dissociating into its component subunits when in common physiological buffers. The conventional method for accomplishing this is through a destructive process wherein the treated APC trimer is chemically disrupted in 8M urea and then allowed to re-associate through in a physiological buffer. An alternative method can be used which preserves the structural integrity of the APC trimer and allows for a brighter, more stable end-product.

PHEOPHYTIN

Pheophytin or phaeophytin (abbreviated Pheo) is a chemical compound that serves as the first electron carrier intermediate in the electron transfer pathway of Photosystem II (PS II) in plants, and the photosynthetic reaction center (RC P870) found in purple bacteria. In both PS II and RC P870, light drives electrons from the reaction center through pheophytin, which then passes the electrons to a quinone (Q_A) in RC P870 and RC P680. The overall mechanisms, roles, and purposes of the pheophytin molecules in the two transport chains are analogous to each other.

Pheophytin a, i.e. chlorophyll a without the Mg^{2+} ion.

Structure

In biochemical terms, pheophytin is a chlorophyll molecule lacking a central Mg^{2+} ion. It can be produced from chlorophyll by treatment with a weak acid, producing a dark bluish waxy pigment. The probable etymology comes from this description, with *pheo* meaning *dusky* and *phyt* meaning *vegetation*.

Reaction in Purple Bacteria

Pheophytin is the first electron carrier intermediate in the photoreaction center (RC P870) of purple bacteria. Its involvement in this system can be broken down into 5 basic steps. The first step is excitation of the bacteriochlorophylls $(Chl)_2$ or the special pair of chlorophylls. This can be seen in the following reaction:

- $(Chl)_2 + 1\,photon \rightarrow (Chl)_2^*$ (excitation)

The second step involves the $(Chl)_2$ passing an electron to pheophytin, producing a negatively charged radical (the pheophytin) and a positively charged radical (the special pair of chlorophylls), which results in a charge separation.

- $(Chl)_2^* + Pheo \rightarrow \cdot(Chl)_2^+ + \cdot Pheo^-$ (charge separation)

The third step is the rapid electron movement to the tightly bound menaquinone, Q_A, which immediately donates the electrons to a second, loosely bound quinone (Q_B). Two electron transfers convert Q_B to its reduced form $(Q_B H_2)$.

- $2 \cdot Pheo^- + 2H^+ + Q_B \rightarrow 2Pheo + Q_B H_2$ (quinone reduction)

The fifth and final step involves the filling of the "hole" in the special pair by an electron from a heme in cytochrome c. This regenerates the substrates and completes the cycle, allowing for subsequent reactions to take place.

Involvement in Photosystem II

In photosystem II, pheophytin plays a very similar role. It again acts as the first electron carrier intermediate in the photosystem. After P680 becomes excited to P680*, it transfers an electron to pheophytin, which converts the molecule into a negatively charged radical. Two negatively charged pheophytin radicals quickly pass their extra electrons to two consecutive plastoquinone molecules. Eventually, the electrons pass through the cytochrome $b_6 f$ molecule and leaves photosystem II. The reactions outlined above in the section concerning purple bacteria give a general illustration of the actual movement of the electrons through pheophytin and the photosystem. The overall scheme is:

- Excitation,

- Charge separation,

- Plastoquinone reduction,

- Regeneration of substrates.

PHYCOBILIN

Phycobilins are water-soluble pigments found in the stroma of chloroplast organelles that are present only in Cyanobacteria and Rhodophyta. The two classes of phycobilins include phycocyanin and phycoerythrin. Phycocyanin is a bluish pigment found in primarily cyanobacteria (blue-green algae) to aid in absorption of light in photosynthesis, while phycoerythrin is a pigment found in Rhodopyta (red algae) that is responsible for its characteristic red color. It is an accessory pigment that allows red algae to carry out photosynthesis in deep water where wavelengths of blue light are most abundant by absorbing blue light and reflecting red light.

Phycocyanin

Phycocyanobilin.

Phycocyanin is a pigment-protein complex from the light-harvesting phycobiliprotein family, along with allophycocyanin and phycoerythrin. It is an accessory pigment to chlorophyll. All phycobiliproteins are water-soluble, so they cannot exist within the membrane like carotenoids can. Instead, phycobiliproteins aggregate to form clusters that adhere to the membrane called phycobilisomes. Phycocyanin is a characteristic light blue color, absorbing orange and red light, particularly near 620 nm (depending on which specific type it is), and emits fluorescence at about 650 nm (also depending on which type it is). Allophycocyanin absorbs and emits at longer wavelengths than phycocyanin C or phycocyanin R. Phycocyanins are found in Cyanobacteria (also called blue-green algae). Phycobiliproteins have fluorescent properties that are used in immunoassay kits. The product phycocyanin, produced by *Aphanizomenon flos-aquae* and Spirulina, is for example used in the food and beverage industry as the natural coloring agent 'Lina Blue' or 'EXBERRY Shade Blue' and is found in sweets and ice cream. In addition, fluorescence detection of phycocyanin pigments in water samples is a useful method to monitor cyanobacteria biomass.

The phycobiliproteins are made of two subunits(alpha and beta) having a protein backbone to which 1-2 linear tetrapyrrole chromophores are covalently bound.

C-phycocyanin is often found in cyanobacteria which thrive around hot springs, as it can be stable up to around 70 °C, with identical spectroscopic (light absorbing) behaviours at 20 and 70 °C. Thermophiles contain slightly different amino acid sequences making it stable under these higher conditions. Molecular weight is around 30,000 Da. Stability of this protein invitro at these temperatures has been shown to be substantially lower. Photo-spectral analysis of the protein after 1 min exposure to 65 °C conditions in a purified state demonstrated a 50% loss of tertiary structure.

Structure

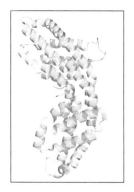

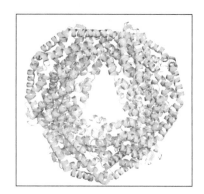

Phycocyanin (αβ) monomer.   Phycocyanin (αβ)$_6$ hexamer.

Phycocyanin shares a common structural theme with all phycobiliproteins. The structure begins with the assembly of phycobiliprotein monomers, which are heterodimers composed of α and β subunits, and their respective chromophores linked via thioether bond.

Each subunit is typically composed of eight α-helices. Monomers spontaneously aggregate to form ring-shaped trimers (αβ)$_3$, which have rotational symmetry and a central channel. Trimers aggregate in pairs to form hexamers (αβ)$_6$, sometimes assisted with additional linker proteins. Each phycobilisome rod generally has two or more phycocyanin hexamers. Despite the overall similarity in structure and assembly of phycobiliproteins, there is a large diversity in hexamer and rod conformations, even when only considering phycocyanins. On a larger scale phycocyanins also vary in crystal structure, although the biological relevance of this is debatable.

As an example, the structure of C-phycocyanin from *Synechococcus vulcanus* has been refined to 1.6 Angstrom resolution. The (αβ) monomer consists of 332 amino acids and 3 thio-linked phycocyanobilin (PCB) cofactor molecules. Both the α- and β-subunits have a PCB at amino acid 84, but the β-subunit has an additional PCB at position 155 as well. This additional PCB faces the exterior of the trimeric ring and is therefore implicated in inter-rod energy transfer in the phycobilisome complex. In addition to cofactors, there are many predictable non-covalent interactions with the surrounding solvent (water) that are hypothesized to contribute to structural stability.

R-phycocyanin II (R-PC II) is found in some *Synechococcus* species. R-PC II is said to be the first PEB containing phycocyanin that originates in cyanobacteria. Its purified protein is composed of alpha and beta subunits in equal quantities. R-PC II has PCB at beta-84 and the phycoerythrobillin (PEB) at alpha-84 and beta-155.

As of March 7, 2018, there are 44 crystal structures of phycocyanin deposited in the Protein Data Bank.

Spectral Characteristics

C-phycocyanin has a single absorption peak at ~621 nm, varying slightly depending on the organism and conditions such as temperature, pH, and protein concentration *in vitro*. Its emission maximum is ~642 nm. This means that the pigment absorbs orange light, and emits reddish light. R-phycocyanin has an absorption maxima at 533 and 544 nm. The fluorescence emission maximum of R-phycocyanin is 646 nm.

Property	C-Phycocyanin	R-Phycocyanin
Absorption maximum (nm)	621	533, 544
Emission maximum (nm)	642	646
Extinction Coefficient (ε)	1.54×10^6 $M^{-1}cm^{-1}$	-
Quantum Yield	0.81	-

Ecological Relevance

Phycocyanin is produced by many photoautotrophic cyanobacteria. Even if cyanobacteria have large concentrations of phycocyanin, productivity in the ocean is still limited due to light conditions.

Phycocyanin has ecological significance in indicating cyanobacteria bloom. Normally chlorophyll *a* is used to indicate cyanobacteria numbers, however since it is present in a large number of phytoplankton groups, it is not an ideal measure. For instance a study in the Baltic Sea used phycocyanin as a marker for filamentous cyanobacteria during toxic summer blooms. Some filamentous organisms in the Baltic Sea include *Nodularia spumigena* and *Aphanizomenon flosaquae*.

An important cyanobacteria named spirulina (*Arthrospira plantensis*) is a micro algae that produces C-PC.

There are many different methods of phycocyanin production including photoautotrophic, mixotrophic and heterotrophic and recombinant production. Photoautotrophic production of phycocyanin is where cultures of cyanobacteria are grown in open ponds in either subtropical or tropical regions. Mixotrophic production of algae is where the algae are grown on cultures that have an organic carbon source like glucose. Using mixotrophic production produces higher growth rates and higher biomass compared to simply using a photoautotrophic culture. In the mixotrophic culture, the sum of heterotrophic and autotrophic growth separately was equal to the mixotrophic growth. Heterotrophic production of phycocyanin is not light limited, as per its definition. *Galdieria sulphuraria* is a unicellular rhodophyte that contains a large amount of C-PC and a small amount of allophycocyanin. *G. sulphuraria* is an example of the heterotrophic production of C-PC because its habitat is hot, acidic springs and uses a number of carbon sources for growth. Recombinant production of C-PC is another heterotrophic method and involves gene engineering.

Lichen-forming fungi and cyanobacteria often have a symbiotic relationship and thus phycocyanin markers can be used to show the ecological distribution of fungi-associated cyanobacteria. As shown in the highly specific association between Lichina species and Rivularia strains, phycocyanin has enough phylogenetic resolution to resolve the evolutionary history of the group across the northwestern Atlantic Ocean coastal margin.

Biosynthesis

The two genes cpcA and cpcB, located in the cpc operon and translated from the same mRNA transcript, encode for the C-PC α- and β-chains respectively. Additional elements such as linker proteins, and enzymes involved in phycobilin synthesis and the phycobiliproteins are often encoded by genes in adjacent gene clusters, and the cpc operon of Arthrospira platensis also encodes a linker protein assisting in the assembly of C-PC complexes. In red algae, the phycobiliprotein and linker protein genes are located on the plastid genome.

Phycocyanobilin is synthesised from heme and inserted into the C-PC apo-protein by three enzymatic steps. Cyclic heme is oxidised to linear biliverdin IXα by heme oxygenase and further converted to 3Z-phycocyanobilin, the dominant phycocyanobilin isomer, by 3Z-phycocyanobilin:ferredoxin oxidoreductase. Insertion of 3Z-phycocyanobilin into the C-PC apo-protein via thioether bond formation is catalysed by phycocyanobilin lyase.

The promoter for the cpc operon is located within the 427-bp upstream region of the cpcB gene. In *A. platensis*, 6 putative promoter sequences have been identified in the region, with four of them showing expression of green fluorescent protein when transformed into *E. coli*. The presence of other positive elements such as light-response elements in the same region have also been demonstrated.

The multiple promoter and response element sequences in the cpc operon enable cyanobacteria and red algae to adjust its expression in response to multiple environmental conditions. Expression of the cpcA and cpcB genes is regulated by light. Low light intensities stimulate synthesis of CPC and other pigments, while pigment synthesis is repressed at high light intensities. Temperature has also been shown to affect synthesis, with specific pigment concentrations showing a clear maximum at 36 °C in Arthronema africanum, a cyanobacterium with particular high C-PC and APC contents.

Nitrogen and also iron limitation induce phycobiliprotein degradation. Organic carbon sources stimulate C-PC synthesis in Anabaena spp., but seem to have almost no effector negative effect in A. platensis. In the rhodophytes Cyanidium caldarium and Galdieria sulphuraria, C-PC production is repressed by glucose but stimulated by heme.

Phycoerythrin

Phycoerythrin (PE) is a red protein-pigment complex from the light-harvesting phycobiliprotein family, present in red algae and cryptophytes, accessory to the main chlorophyll pigments responsible for photosynthesis.

Like all phycobiliproteins, it is composed of a protein part covalently binding chromophores called phycobilins. In the phycoerythrin family, the most known phycobilins are: phycoerythrobilin, the typical phycoerythrin acceptor chromophore, and sometimes phycourobilin. Phycoerythrins are composed of (αβ) monomers, usually organised in a disk-shaped trimer (αβ)$_3$ or hexamer (αβ)$_6$ (second one is the functional unit of the antenna rods). These typical complexes also contain a third type of subunit, the γ chain.

Phycobilisomes

Phycobiliproteins are part of huge light harvesting antennae protein complexes called phycobilisomes. In red algae they are anchored to the stromal side of thylakoid membranes of chloroplasts, whereas in cryptophytes phycobilisomes are reduced and (phycobiliprotein 545 PE545 molecules here) are densely packed inside the lumen of thylakoides.

Phycoerythrin is an accessory pigment to the main chlorophyll pigments responsible for photosynthesis. The light energy is captured by phycoerythrin and is then passed on to the reaction centre chlorophyll pair, most of the time via the phycobiliproteins phycocyanin and allophycocyanin.

Structural Characteristics

Phycoerythrins except phycoerythrin 545 (PE545) are composed of (αβ) monomers assembled into disc-shaped (αβ)$_6$ hexamers or (αβ)$_3$ trimers with 32 or 3 symmetry and enclosing central channel. In phycobilisomes (PBS) each trimer or hexamer contains at least one linker protein located in central channel. B-phycoerythrin (B-PE) and R-phycoerythrin (R-PE) from red algae in addition to α and β chains have a third, γ subunit contributing both linker and light-harvesting functions, because it bears chromophores.

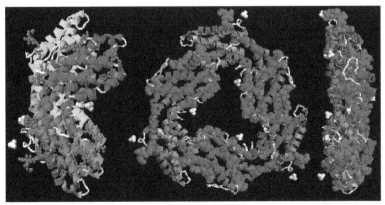

The crystal structure of B-phycoerythrin from red algae *Porphyridium cruentum* (PDB ID: 3V57).
The asymmetric unit (αβ)$_2$ on the left and assumed biological molecule (αβ)$_3$.
It contains phycoerythrobilin, N-methyl asparagine and SO_4^{2-}.

R-phycoerythrin is predominantly produced by red algae. The protein is made up of at least three different subunits and varies according to the species of algae that produces it. The subunit structure of the most common R-PE is $(\alpha\beta)_6\gamma$. The α subunit has two phycoerythrobilins (PEB), the β subunit has 2 or 3 PEBs and one phycourobilin (PUB), while the different gamma subunits are reported to have 3 PEB and 2 PUB (γ_1) or 1 or 2 PEB and 1 PUB (γ_2). The molecular weight of R-PE is 250,000 Daltons.

Crystal structures available in the Protein Data Bank contain in one $(\alpha\beta)_2$ or $(\alpha\beta\gamma)_2$ asymmetric unit of different phycoerythrins:

Phycoerythrobilin is the typical chromophore in phycoerythrin. It is similar to porphyrin of chlorophyll for example, but tetrapyrrole is linear, not closed into ring with metal ion in the middle.

The red algae *Gracilaria* contains R-phycoerythrin.

Chromophore or other non-protein molecule	Phycoerythrin				Chain
	PE545	B-PE	R-PE	Other types	
Bilins	8	10	10	10	α and β
- Phycoerythrobilin (PEB)	6	10	0 or 8	8	β (PE545) or α and β
- 15,16-dihydrobiliverdin (DBV)	2	-	-	-	α (-3 and -2)
- Phycocyanobilin (CYC)	-	-	8 or 7 or 0	-	α and β
- Biliverdine IX alpha (BLA)	-	-	0 or 1	-	α
- Phycourobilin (PUB)	-	-	2	2	β
5-hydroxylysine (LYZ)	1 or 2	-	-	-	α (-3 or -3 and -2)

N-methyl asparagine (MEN)	2	2	0 or 2	2	β
Sulfate ion SO$_4^{2-}$ (SO4)	-	5 or 1	0 or 2	-	α or α and β
Chloride ion Cl$^-$ (CL)	1	-	-	-	β
Magnesium ion Mg^{2+} (MG)	2	-	-	-	α-3 and β
inspected PDB files	1XG0 1XF6 1QGW	3V57 3V58	1EYX 1LIA 1B8D	2VJH	

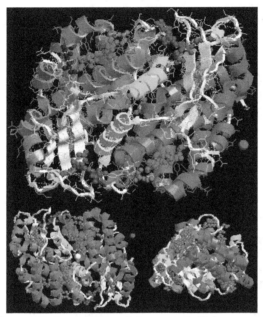

The crystal structure of phycoerythrin 545 (PE545) from a unicellular cryptophyte *Rhodomonas* CS24 (PDB ID: 1XG0). Colors: chains - alpha-2, alpha-3, beta, beta (helixes, sheets are yellow), phycoerythrobilin, 15,16-dihydrobiliverdin (15,16-DHBV), 5-hydroxylysine, N-methyl asparagine, Mg^{2+}, Cl$^-$.

The assumed biological molecule of phycoerythrin 545 (PE545) is $(\alpha\beta)_2$ or rather $(\alpha_3\beta)$ $(\alpha_2\beta)$. The numbers 2 and 3 after the α letters in second formula are part of chain names here, not their counts. The synonym cryptophytan name of α_3 chain is α_1 chain.

The largest assembly of B-phycoerythrin (B-PE) is $(\alpha\beta)_3$ trimer . However, preparations from red algae yield also $(\alpha\beta)_6$ hexamer. In case of R-phycoerythrin (R-PE) the largest assumed biological molecule here is $(\alpha\beta\gamma)_6$, $(\alpha\beta\gamma)_3(\alpha\beta)_3$ or $(\alpha\beta)_6$ dependently on publication, for other phycoerythrin types $(\alpha\beta)_6$. These γ chains from the Protein Data Bank are very small and consist only of three or six recognizable amino acids, whereas described at the beginning of this section linker γ chain is large (for example 277 amino acid long 33 kDa in case of γ^{33} from red algae *Aglaothamnion neglectum*). This is because the electron density of the gamma-polypeptide is mostly averaged out by its threefold crystallographic symmetry and only a few amino acids can be modeled.

For $(\alpha\beta\gamma)_6$, $(\alpha\beta)_6$ or $(\alpha\beta\gamma)_3(\alpha\beta)_3$ the values from the table should be simply multiplied by 3, $(\alpha\beta)_3$ contain intermediate numbers of non-protein molecules.

In phycoerythrin PE545 above, one α chain (-2 or -3) binds one molecule of billin, in other examples it binds two molecules. The β chain always binds to three molecules. The small γ chain binds to none.

Two molecules of N-methyl asparagine are bound to the β chain, one 5-hydroxylysine to α (-3 or -2), one Mg^{2+} to α-3 and β, one Cl^- to β, 1-2 molecules of SO_4^{2-} to α or β.

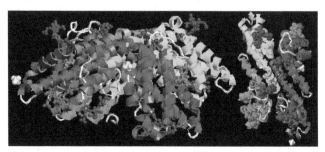

The crystal structure of R-phycoerythrin from red algae *Gracilaria chilensis* (PDB ID: 1EYX) - basic oligomer $(αβγ)_2$ (so called asymmetric unit). It contains phycocyanobilin, biliverdine IX alpha, phycourobilin, N-methyl asparagine, SO_4^{2-}. One fragment of γ chain is red, second one white because it is not considered as alpha helix despite identical aminoacid sequence.

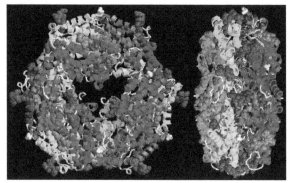

The entire oligomer of R-phycoerythrin from *Gracilaria chilensis* $(αβγ)_6$ (PDB ID: 1EYX).

Spectral Characteristics

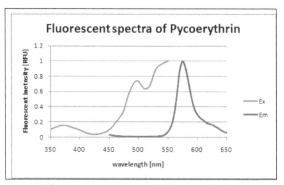

Fluorescent spectra of phycoerythrin.

Absorption peaks in the visible light spectrum are measured at 495 and 545/566 nm, depending on the chromophores bound and the considered organism. A strong emission peak exists at 575 ± 10 nm. (*i.e.*, phycoerythrin absorbs slightly blue-green/yellowish light and emits slightly orange-yellow light.)

Property	Value
Absorption maximum	565 nm
Additional Absorption peak	498 nm
Emission maximum	573 nm
Extinction Coefficient (ε)	$1.96 \times 10^6 \ M^{-1}cm^{-1}$
Quantum Yield (QY)	0.84
Brightness ($\varepsilon \times$ QY)	$1.65 \times 10^6 \ M^{-1}cm^{-1}$

PEB and DBV bilins in PE545 absorb in the green spectral region too, with maxima at 545 and 569 nm respectively. The fluorescence emission maximum is at 580 nm.

R-Phycoerythrin Variations

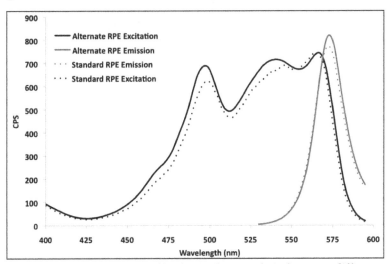

Excitation and emission profiles for R-Phycoerythrin from two different algae. Common laser excitation wavelengths are also noted.

As mentioned above, phycoerythrin can be found in a variety of algal species. As such, there can be variation in the efficiency of absorbance and emission of light required for facilitation of photosynthesis. This could be a result of the depth in the water column that a specific alga typically resides and a consequent need for greater or less efficiency of the accessory pigments.

With advances in imaging and detection technology which can avoid rapid photobleaching, protein fluorophores have become a viable and powerful tool for researchers in fields such as microscopy, microarray analysis and Western blotting. In light of this, it may be beneficial for researchers to screen these variable R-phycoerythrins to determine which one is most appropriate for their particular application. Even a small increase in fluorescent efficiency could reduce background noise and lower the rate of false-negative results.

RUBISCO

Ribulose-1,5-bisphosphate carboxylase/oxygenase, commonly known by the abbreviations Rubisco, rubisco, RuBPCase, or RuBPco, is an enzyme involved in the first major step of carbon fixation, a process by which atmospheric carbon dioxide is converted by plants and other photosynthetic organisms to energy-rich molecules such as glucose. In chemical terms, it catalyzes the carboxylation of ribulose-1,5-bisphosphate (also known as *RuBP*). It is probably the most abundant enzyme on Earth.

Alternative Carbon Fixation Pathways

RuBisCO is important biologically because it catalyzes the primary chemical reaction by which inorganic carbon enters the biosphere. While many autotrophic bacteria and archaea fix carbon via the reductive acetyl CoA pathway, the 3-hydroxypropionate cycle, or the reverse Krebs cycle, these pathways are relatively small contributors to global carbon fixation than that catalyzed by RuBisCO. Phosphoenolpyruvate carboxylase, unlike RuBisCO, only temporarily fixes carbon. Reflecting its importance, RuBisCO is the most abundant protein in leaves, accounting for 50% of soluble leaf protein in C_3 plants (20–30% of total leaf nitrogen) and 30% of soluble leaf protein in C_4 plants (5–9% of total leaf nitrogen). Given its important role in the biosphere, the genetic engineering of RuBisCO in crops is of continuing interest.

Structure

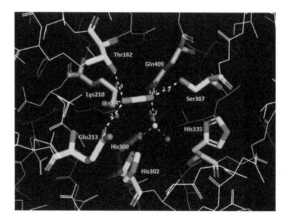

Active site of RuBisCO of *Galdieria sulphuraria* with CO_2. Residues involved in both the active site and stabilizing CO_2 for enzyme catalysis are shown in color and labeled.

Distances of the hydrogen bonding interactions are shown in Angstroms. Mg^{2+} ion (green sphere) is shown coordinated to CO_2, and is followed by three water molecules (red spheres). All other residues are placed in grayscale.

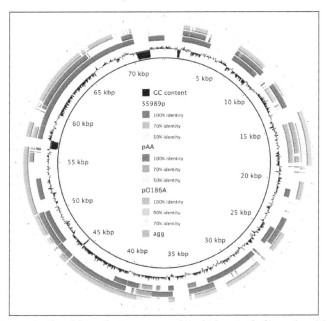

Location of the rbcL gene in the chloroplast genome of Arabidopsis thaliana (positions ca. 55-56.4 kb). rbcL is one of the 21 protein-coding genes involved in photosynthesis (green boxes).

In plants, algae, cyanobacteria, and phototrophic and chemoautotrophic proteobacteria, the enzyme usually consists of two types of protein subunit, called the large chain (L, about 55,000 Da) and the small chain (S, about 13,000 Da). The *large-chain* gene (*rbcL*) is encoded by the chloroplast DNA in plants. There are typically several related *small-chain* genes in the nucleus of plant cells, and the small chains are imported to the stromal compartment of chloroplasts from the cytosol by crossing the outer chloroplast membrane. The enzymatically active substrate (ribulose 1,5-bisphosphate) binding sites are located in the large chains that form dimers in which amino acids from each large chain contribute to the binding sites. A total of eight large-chains (= 4 dimers) and eight small chains assemble into a larger complex of about 540,000 Da. In some proteobacteria and dinoflagellates, enzymes consisting of only large subunits have been found.

Magnesium ions (Mg^{2+}) are needed for enzymatic activity. Correct positioning of Mg^{2+} in the active site of the enzyme involves addition of an "activating" carbon dioxide molecule (CO_2) to a lysine in the active site (forming a carbamate). Mg^{2+} operates by driving deprotonation of the Lys210 residue, causing the Lys residue to rotate by 120 degrees to the *trans* conformer, decreasing the distance between the nitrogen of Lys and the carbon of CO_2. The close proximity allows for the formation of a covalent bond, resulting in the carbamate. Mg^{2+} is first enabled to bind to the active site by the rotation of His335 to an alternate conformation. Mg^{2+} is then coordinated by the His residues of the active site (His300, His302, His335), and is partially neutralized by the coordination of three water molecules and their conversion to $^-$OH. This coordination results in an unstable complex, but produces a favorable environment for the binding of Mg^{2+}. Formation of the carbamate is favored by an alkaline pH. The pH and the concentration of magnesium ions in the fluid compartment (in plants, the stroma of the chloroplast) increases in the light. The role of changing pH and magnesium ion levels in the regulation of RuBisCO enzyme activity is discussed below. Once the carbamate is formed, His335 finalizes the activation by returning to its initial position through thermal fluctuation.

RuBisCO large chain, catalytic domain	
Identifiers	
Symbol	RuBisCO_large
Pfam	PF00016
InterPro	IPR000685
PROSITE	PDOC00142
SCOPe	3rub / SUPFAM
CDD	cd08148
showAvailable protein structures:	
Pfam	structures / ECOD
PDB	RCSB PDB; PDBe; PDBj
PDBsum	structure summary
PDB	1aa1, 1aus, 1bwv, 1bxn, 1ej7, 1geh, 1gk8, 1ir1, 1ir2, 1iwa, 1rba, 1rbl, 1rbo, 1rco, 1rcx, 1rld, 1rsc, 1rus, 1rxo, 1svd, 1tel, 1upm, 1upp, 1uw9, 1uwa, 1uzd, 1uzh, 1wdd, 1ykw, 2cwx, 2cxe, 2d69, 2qyg, 2rus, 2v63, 2v67, 2v68, 2v69, 2v6a, 3rub, 4rub, 5rub, 8ruc, 9rub

RuBisCO, N-terminal domain	
Identifiers	
Symbol	RuBisCO_large_N
Pfam	PF02788
InterPro	IPR017444
SCOPe	3rub / SUPFAM
showAvailable protein structures:	
Pfam	structures / ECOD
PDB	RCSB PDB; PDBe; PDBj
PDBsum	structure summary
PDB	1aa1, 1aus, 1bwv, 1bxn, 1ej7, 1geh, 1gk8, 1ir1, 1ir2, 1iwa, 1rba, 1rbl, 1rbo, 1rco, 1rcx, 1rld, 1rsc, 1rus, 1rxo, 1svd, 1tel, 1upm, 1upp, 1uw9, 1uwa, 1uzd, 1uzh, 1wdd, 1ykw, 2cwx, 2cxe, 2d69, 2qyg, 2rus, 2v63, 2v67, 2v68, 2v69, 2v6a, 3rub, 4rub, 5rub, 8ruc, 9rub

RuBisCO, small chain	
Identifiers	
Symbol	RuBisCO_small
Pfam	PF00101
InterPro	IPR000894
SCOPe	3rub / SUPFAM
CDD	cd03527
showAvailable protein structures:	
Pfam	structures / ECOD
PDB	RCSB PDB; PDBe; PDBj
PDBsum	structure summary
PDB	1aa1, 1aus, 1bwv, 1bxn, 1ej7, 1gk8, 1ir1, 1ir2, 1iwa, 1rbl, 1rbo, 1rco, 1rcx, 1rlc, 1rld, 1rsc, 1rxo, 1svd, 1upm, 1upp, 1uw9, 1uwa, 1uzd, 1uzh, 1wdd, 2v63, 2v67, 2v68, 2v69, 2v6a, 3rub, 4rub, 8ruc

Enzymatic Activity

Two main reactions of RuBisCo: CO_2 fixation and oxygenation.

RuBisCO is one of many enzymes in the Calvin cycle. When Rubisco facilitates the attack of CO_2 at the C_2 carbon of RuBP and subsequent bond cleavage between the C_3 and C_2 carbon, 2 molecules of glycerate-3-phosphate are formed. The conversion involves these steps: enolisation, carboxylation, hydration, C-C bond cleavage, and protonation.

Substrates

Substrates for RuBisCO are ribulose-1,5-bisphosphate and carbon dioxide (distinct from the "activating" carbon dioxide). RuBisCO also catalyses a reaction of ribulose-1,5-bisphosphate and molecular oxygen (O_2) instead of carbon dioxide (CO_2). Discriminating between the substrates CO_2 and O_2 is attributed to the differing interactions of the substrate's quadrupole moments and a high electrostatic field gradient. This gradient is established by the dimer form of the minimally active RuBisCO, which with its two components provides a combination of oppositely charged domains required for the enzyme's interaction with O_2 and CO_2. These conditions help explain the low turnover rate found in RuBisCO: In order to increase the strength of the electric field necessary for sufficient interaction with the substrates' quadrupole moments, the C- and N- terminal segments of the enzyme must be closed off, allowing the active site to be isolated from the solvent and lowering the dielectric constant. This isolation has a significant entropic cost, and results in the poor turnover rate.

Binding RuBP

Carbamylation of the ε-amino group of Lys201 is stabilized by coordination with the Mg^{2+}. This reaction involves binding of the carboxylate termini of Asp203 and Glu204 to the Mg^{2+} ion. The substrate RuBP binds Mg^{2+} displacing two of the three aquo ligands.

Enolisation

Enolisation of RuBP is the conversion of the keto tautomer of RuBP to an enediol(ate). Enolisation is initiated by deprotonation at C_3. The enzyme base in this step has been debated, but the steric constraints observed in crystal structures have made Lys201 the most likely candidate. Specifically, the carbamate oxygen on Lys201 that is not coordinated with the Mg ion deprotonates the C_3 carbon of RuBP to form a 2,3-enediolate.

Carboxylation

A 3D image of the active site of spinach RuBisCO complexed with the inhibitor 2-Carboxyarabinitol-1,5-Bisphosphate, CO_2, and Mg_2^+. (PDB: 1IR1; Ligand View [CAP]501:A).

Carboxylation of the 2,3-enediolate results in the intermediate 3-keto-2'-carboxyarabinitol-1,5-bisphosphate and Lys334 is positioned to facilitate the addition of the CO_2 substrate as it replaces the third Mg_2^+-coordinated water molecule and add directly to the enediol. No Michaelis complex is formed in this process. Hydration of this ketone results in an additional hydroxy group on C_3, forming a gem-diol intermediate. Carboxylation and hydration have been proposed as either a single concerted step or as two sequential steps. Concerted mechanism is supported by the proximity of the water molecule to C_3 of RuBP in multiple crystal structures. Within the spinach structure, other residues are well placed to aid in the hydration step as they are within hydrogen bonding distance of the water molecule.

C-C bond Cleavage

The gem-diol intermediate cleaves at the C_2-C_3 bond to form one molecule of glycerate-3-phosphate and a negatively charge carboxylate. Stereo specific protonation of C_2 of this carbanion results in another molecule of glycerate-3-phosphate. This step is thought to be facilitated by Lys175 or potentially the carbamylated Lys201.

Products

When carbon dioxide is the substrate, the product of the carboxylase reaction is a unstable six-carbon phosphorylated intermediate known as 3-keto-2-carboxyarabinitol-1,5-bisphosphate, which decays rapidly into two molecules of glycerate-3-phosphate. The 3-phosphoglycerate can be used to produce larger molecules such as glucose.

Rubisco side activities can lead to useless or inhibitory by-products; one such product is xylulose-1,5-bisphosphate, which inhibits Rubisco activity.

When molecular oxygen is the substrate, the products of the oxygenase reaction are phosphoglycolate and 3-phosphoglycerate. Phosphoglycolate is recycled through a sequence of reactions called photorespiration, which involves enzymes and cytochromes located in the mitochondria and peroxisomes (this is a case of metabolite repair). In this process, two molecules of phosphoglycolate are converted to one molecule of carbon dioxide and one molecule of 3-phosphoglycerate, which can reenter the Calvin cycle. Some of the phosphoglycolate entering this pathway can be retained by plants to produce other molecules such as glycine. At ambient levels of carbon dioxide and oxygen, the ratio of the reactions is about 4 to 1, which results in a net carbon dioxide fixation of only 3.5. Thus, the inability of the enzyme to prevent the reaction with oxygen greatly reduces the photosynthetic capacity of many plants. Some plants, many algae, and photosynthetic bacteria have overcome this limitation by devising means to increase the concentration of carbon dioxide around the enzyme, including C_4 carbon fixation, crassulacean acid metabolism, and the use of pyrenoid.

Rate of Enzymatic Activity

Some enzymes can carry out thousands of chemical reactions each second. However, RuBisCO is slow, fixing only 3-10 carbon dioxide molecules each second per molecule of enzyme. The reaction catalyzed by RuBisCO is, thus, the primary rate-limiting factor of the Calvin cycle during the day. Nevertheless, under most conditions, and when light is not otherwise limiting photosynthesis, the speed of RuBisCO responds positively to increasing carbon dioxide concentration.

RuBisCO is usually only active during the day as ribulose 1,5-bisphosphate is not regenerated in the dark. This is due to the regulation of several other enzymes in the Calvin cycle. In addition, the activity of RuBisCO is coordinated with that of the other enzymes of the Calvin cycle in several ways.

By Ions

Upon illumination of the chloroplasts, the pH of the stroma rises from 7.0 to 8.0 because of the proton (hydrogen ion, H^+) gradient created across the thylakoid membrane. The movement of protons into thylakoids is driven by light and is fundamental to ATP synthesis in chloroplasts. At the same time, magnesium ions (Mg^{2+}) move out of the thylakoids, increasing the concentration of magnesium in the stroma of the chloroplasts. RuBisCO has a high optimal pH (can be >9.0, depending on the magnesium ion concentration) and, thus, becomes "activated" by the addition of carbon dioxide and magnesium to the active sites as described above.

By RuBisCO Activase

In plants and some algae, another enzyme, RuBisCO activase, is required to allow the rapid formation of the critical carbamate in the active site of RuBisCO. RuBisCO activase is required because the ribulose 1,5-bisphosphate (RuBP) substrate binds more strongly to the active sites lacking the carbamate and markedly slows down the "activation" process. In the light, RuBisCO activase promotes the release of the inhibitory (or — in some views — storage) RuBP from the catalytic sites. Activase is also required in some plants (e.g., tobacco and many beans) because, in darkness, RuBisCO is inhibited (or protected from hydrolysis) by a competitive inhibitor synthesized by these plants, a substrate analog 2-Carboxy-D-arabitinol 1-phosphate (CA1P). CA1P binds tightly to the active site of carbamylated RuBisCO and inhibits catalytic activity. In the light, RuBisCO activase also promotes the release of CA1P from the catalytic sites. After the CA1P is released from RuBisCO, it is rapidly converted to a non-inhibitory form by a light-activated CA1P-phosphatase. Finally, once every several hundred reactions, the normal reactions with carbon dioxide or oxygen are not completed, and other inhibitory substrate analogs are formed in the active site. Once again, RuBisCO activase can promote the release of these analogs from the catalytic sites and maintain the enzyme in a catalytically active form. The properties of activase limit the photosynthetic potential of plants at high temperatures. CA1P has also been shown to keep RuBisCO in a conformation that is protected from proteolysis. At high temperatures, RuBisCO activase aggregates and can no longer activate RuBisCO. This contributes to the decreased carboxylating capacity observed during heat stress.

By ATP/ADP and Stromal Reduction/Oxidation State through the Activase

The removal of the inhibitory RuBP, CA1P, and the other inhibitory substrate analogs by activase requires the consumption of ATP. This reaction is inhibited by the presence of ADP, and, thus, activase activity depends on the ratio of these compounds in the chloroplast stroma. Furthermore, in most plants, the sensitivity of activase to the ratio of ATP/ADP is modified by the stromal reduction/oxidation (redox) state through another small regulatory protein, thioredoxin. In this manner, the activity of activase and the activation state of RuBisCO can be modulated in response to light intensity and, thus, the rate of formation of the ribulose 1,5-bisphosphate substrate.

By Phosphate

In cyanobacteria, inorganic phosphate (P_i) participates in the co-ordinated regulation of photosynthesis. P_i binds to the RuBisCO active site and to another site on the large chain where it can influence transitions between activated and less active conformations of the enzyme. Activation of bacterial RuBisCO might be particularly sensitive to P_i levels, which can act in the same way as RuBisCO activase in higher plants.

By Carbon Dioxide

Since carbon dioxide and oxygen compete at the active site of RuBisCO, carbon fixation by RuBisCO can be enhanced by increasing the carbon dioxide level in the compartment containing RuBisCO (chloroplast stroma). Several times during the evolution of plants, mechanisms have evolved for increasing the level of carbon dioxide in the stroma. The use of oxygen as a substrate appears to be a puzzling process, since it seems to throw away captured energy. However, it may be a mechanism for preventing overload during periods of high light flux. This weakness in the enzyme is the cause of photorespiration, such that healthy leaves in bright light may have zero net carbon fixation when the ratio of O_2 to CO_2 reaches a threshold at which oxygen is fixed instead of carbon. This phenomenon is primarily temperature-dependent. High temperature decreases the concentration of CO_2 dissolved in the moisture in the leaf tissues. This phenomenon is also related to water stress. Since plant leaves are evaporatively cooled, limited water causes high leaf temperatures. C_4 plants use the enzyme PEP carboxylase initially, which has a higher affinity for CO_2. The process first makes a 4-carbon intermediate compound, which is shuttled into a site of C_3 photosynthesis then de-carboxylated, releasing CO_2 to boost the concentration of CO_2, hence the name C_4 plants.

Crassulacean acid metabolism (CAM) plants keep their stomata closed during the day, which conserves water but prevents the light-independent reactions (a.k.a. the Calvin Cycle) from taking place, since these reactions require CO_2 to pass by gas exchange through these openings. Evaporation through the upper side of a leaf is prevented by a layer of wax.

Genetic Engineering

Since RuBisCO is often rate-limiting for photosynthesis in plants, it may be possible to improve photosynthetic efficiency by modifying RuBisCO genes in plants to increase catalytic activity and/or decrease oxygenation rates. This could improve biosequestration of CO_2 and be both an important climate change strategy and a strategy to increase crop yields. Approaches under investigation include transferring RuBisCO genes from one organism into another organism, engineering Rubisco activase from thermophilic cyanobacteria into temperature sensitive plants, increasing the level of expression of RuBisCO subunits, expressing RuBisCO small chains from the chloroplast DNA, and altering RuBisCO genes to increase specificity for carbon dioxide or otherwise increase the rate of carbon fixation.

Mutagenesis in Plants

In general, site-directed mutagenesis of RuBisCO has been mostly unsuccessful, though mutated forms of the protein have been achieved in tobacco plants with subunit C_4 species, and a RuBisCO with more C_4-like kinetic characteristics have been attained in rice via nuclear transformation.

One avenue is to introduce RuBisCO variants with naturally high specificity values such as the ones from the red alga *Galdieria partita* into plants. This may improve the photosynthetic efficiency of crop plants, although possible negative impacts have yet to be studied. Advances in this area include the replacement of the tobacco enzyme with that of the purple photosynthetic bacterium *Rhodospirillum rubrum*. In 2014, two transplastomic tobacco lines with functional RuBisCO from the cyanobacterium *Synechococcus elongatus* PCC7942 (Se7942) were created by replacing the RuBisCO with the large and small subunit genes of the Se7942 enzyme, in combination with either the corresponding Se7942 assembly chaperone, RbcX, or an internal carboxysomal protein, CcmM35. Both mutants had increased CO2 fixation rates when measured as carbon molecules per RuBisCO. However, the mutant plants grew more slowly than wild-type.

A recent theory explores the trade-off between the relative specificity (i.e., ability to favour CO2 fixation over O_2 incorporation, which leads to the energy-wasteful process of photorespiration) and the rate at which product is formed. The authors conclude that RuBisCO may actually have evolved to reach a point of 'near-perfection' in many plants (with widely varying substrate availabilities and environmental conditions), reaching a compromise between specificity and reaction rate. It has been also suggested that the oxygenase reaction of RuBisCO prevents CO_2 depletion near its active sites and provides the maintenance of the chloroplast redox state.

Since photosynthesis is the single most effective natural regulator of carbon dioxide in the Earth's atmosphere, a biochemical model of RuBisCO reaction is used as the core module of climate change models. Thus, a correct model of this reaction is essential to the basic understanding of the relations and interactions of environmental models.

Expression in Bacterial Hosts

There currently are very few effective methods for expressing functional plant RuBisCO in bacterial hosts for genetic manipulation studies. This is largely due to RuBisCO's requirement of complex cellular machinery for its biogenesis and metabolic maintenance including the nuclear-encoded RbcS subunits, which are typically imported into chloroplasts as unfolded proteins. Furthermore, sufficient expression and interaction with RuBisCO activase are major challenges as well. One successful method for expression of RuBisCO in E. coli involves the coexpression of multiple chloroplast chaperones, though this has only been shown in *Arabidopsis thaliana RuBisCO*.

PLASTID TERMINAL OXIDASE

Plastid terminal oxidase or plastoquinol terminal oxidase (PTOX) is an enzyme that resides on the thylakoid membranes of plant and algae chloroplasts and on the membranes of cyanobacteria. The enzyme was hypothesized to exist as a photosynthetic oxidase in 1982 and was verified by sequence similarity to the mitochondrial alternative oxidase (AOX). The two oxidases evolved from a common ancestral protein in prokaryotes, and they are so functionally and structurally similar that a thylakoid-localized AOX can restore the function of a PTOX knockout.

Function

Plastid terminal oxidase catalyzes the oxidation of the plastoquinone pool, which exerts a variety of effects on the development and functioning of plant chloroplasts.

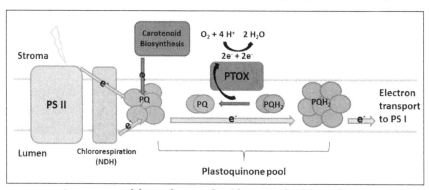

A summary of the pathways plastid terminal oxidase plays a
role in through oxidation of the quinone pool.

Carotenoid Biosynthesis and Plastid Development

The enzyme is important for carotenoid biosynthesis during chloroplast biogenesis. In developing plastids, its activity prevents the over-reduction of the plastoquinone pool. Knockout plants for PTOX exhibit phenotypes of variegated leaves with white patches. Without the enzyme, the carotenoid synthesis pathway slows down due to the lack of oxidized plastoquinone with which to oxidize phytoene, a carotenoid intermediate. The colorless compound phytoene accumulates in the leaves, resulting in white patches of cells. PTOX is also thought to determine the redox poise of the developing photosynthetic apparatus and without it, plants fail to assemble organized internal membrane structures in chloroplasts when exposed to high light during early development.

Photoprotection

Plants deficient in the *IMMUTANS* gene that encodes the oxidase are especially susceptible to photooxidative stress during early plastid development. The knockout plants exhibit a phenotype of variegated leaves with white patches that indicate a lack of pigmentation or photodamage. This effect is enhanced with increased light and temperature during plant development. The lack of plastid terminal oxidase indirectly causes photodamage during plastid development because protective carotenoids are not synthesized without the oxidase.

The enzyme is also thought to act as a safety valve for stress conditions in the photosynthetic apparatus. By providing an electron sink when the plastoquinone pool is over-reduced, the oxidase is thought to protect photosystem II from oxidative damage. Knockouts for Rubisco and photosystem II complexes, which would experience more photodamage than normal, exhibit an upregulation of plastid terminal oxidase. This effect is not universal because it requires plants to have additional PTOX regulation mechanisms. While many studies agree with the stress-protective role of the enzyme, one study showed that overexpression of *PTOX* increases the production of reactive oxygen species and causes more photodamage than normal. This finding suggests that an efficient antioxidant system is required for the oxidase to function as a safety valve for stress conditions and that it is more important during chloroplast biogenesis than in the regular functioning of the chloroplast.

Chlororespiration and Electron Flux

The most confirmed function of plastid terminal oxidase in developed chloroplasts is its role in chlororespiration. In this process, NADPH dehydrogenase (NDH) reduces the quinone pool and the terminal oxidase oxidizes it, serving the same function as cytochrome c oxidase from mitochondrial electron transport. In *Chlamydomonas*, there are two copies of the gene for the oxidase. *PTOX2* significantly contributes to the flux of electrons through chlororespiration in the dark. There is also evidence from experiments with tobacco that it functions in plant chlororespiration as well.

In fully developed chloroplasts, prolonged exposure to light increases the activity of the oxidase. Because the enzyme acts at the plastoquinone pool in between photosystem II and photosystem I, it may play a role in controlling electron flow through photosynthesis by acting as an alternative electron sink. Similar to its role in carotenoid synthesis, its oxidase activity may prevent the over-reduction of photosystem I electron acceptors and damage by photoinhibition. A recent analysis of electron flux through the photosynthetic pathway shows that even when activated, the electron flux plastid terminal oxidase diverts is two orders of magnitude less than the total flux through photosynthetic electron transport. This suggests that the protein may play less of a role than previously thought in relieving the oxidative stress in photosynthesis.

Structure

Plastid terminal oxidase is an integral membrane protein, or more specifically, an integral monotopic protein and is bound to the thylakoid membrane facing the stroma. Based on sequence homology, the enzyme is predicted to contain four alpha helix domains that encapsulate a di-iron center. The two iron atoms are ligated by six essential conserved histidine and glutamate residues – Glu136, Glu175, His171, Glu227, Glu296, and His299. The predicted structure is similar to that of the alternative oxidase, with an additional Exon 8 domain that is required for the plastid oxidase's activity and stability. The enzyme is anchored to the membrane by a short fifth alpha helix that contains a Tyr212 residue hypothesized to be involved in substrate binding.

Mechanism

The oxidase catalyzes the transfer of four electrons from reduced plastoquinone to molecular oxygen to form water. The net reaction is written below:

$$2\,QH_2 + O_2 \rightarrow 2\,Q + 2\,H_2O$$

Analysis of substrate specificity revealed that the enzyme almost exclusively catalyzes the reduction of plastoquinone over other quinones such as ubiquinone and duroquinone. Additionally, iron is essential for the catalytic function of the enzyme and cannot be substituted by another metal cation like Cu^{2+}, Zn^{2+}, or Mn^{2+} at the catalytic center.

It is unlikely that four electrons could be transferred at once in a single iron cluster, so all of the proposed mechanisms involve two separate two-electron transfers from reduced plastoquinone to the di-iron center. In the first step common to all proposed mechanisms, one plastoquinone is oxidized and both irons are reduced from iron(III) to iron(II). Four different mechanisms are proposed for the next step, oxygen capture. One mechanism proposes a peroxide intermediate, after which one oxygen atom is used to create water and another is left bound in a diferryl configuration. Upon one more plastoquinone oxidation, a second water molecule is formed and the irons return to a +3 oxidation state. The other mechanisms involve the formation of Fe(III)-OH or Fe(IV)-OH and a tyrosine radical. These radical-based mechanisms could explain why over-expression of the *PTOX* gene causes increased generation of reactive oxygen species.

Evolution

The enzyme is present in organisms capable of oxygenic photosynthesis, which includes plants, algae, and cyanobacteria. Plastid terminal oxidase and alternative oxidase are thought to have originated from a common ancestral di-iron carboxylate protein. Oxygen reductase activity was likely an ancient mechanism to scavenge oxygen in the early transition from an anaerobic to aerobic world. The plastid oxidase first evolved in ancient cyanobacteria and the alternative oxidase in proteobacteria before eukaryotic evolution and endosymbiosis events. Through endosymbiosis, the plastid oxidase was vertically inherited by eukaryotes that evolved into plants and algae. Sequenced genomes of various plant and algae species shows that the amino acid sequence is more than 25% conserved, which is a significant amount of conservation for an oxidase. This sequence conservation further supports the theory that both the alternative and plastid oxidases evolved before endosymbiosis and did not significantly change through eukaryote evolution.

There also exist PTOX cyanophages that contain copies of the gene for the plastid oxidase. They are known to act as viral vectors for movement of the gene between cyanobacterial species. Some evidence suggests that the phages may use the oxidase to influence photosynthetic electron flow to produce more ATP and less NADPH because viral synthesis utilizes more ATP.

References

- Sun X, Wen T (December 2011). "Physiological roles of plastid terminal oxidase in plant stress responses". J. Biosci. 36(5): 951–6. Doi:10.1007/s12038-011-9161-7. PMID 22116293

- Chlorophyll, entry: newworldencyclopedia.org, Retrieved 6 April, 2019

- Wine & Spirits Education Trust "Wine and Spirits: Understanding Wine Quality" pg 16, Second Revised Edition (2012), London, ISBN 9781905819157

- Chlorophyll-definition-role-in-photosynthesis-4117432: thoughtco.com, Retrieved 7 May, 2019

- Phyt. Collinsdictionary.com. Collins English Dictionary - Complete & Unabridged 11th Edition. Retrieved October 19, 2012

- Chlorophyll, entry: newworldencyclopedia.org, Retrieved 8 June, 2019

- Kubis S, Patel R, Combe J, et al. (August 2004). "Functional specialization amongst the Arabidopsis Toc159 family of chloroplast protein import receptors". Plant Cell. 16 (8): 2059–77. Doi:10.1105/tpc.104.023309. PMC 519198. PMID 15273297

- Bacteriochlorophyll, dictionaries-thesauruses-pictures-and-press-releases, science: encyclopedia.com, Retrieved 9 July, 2019

- "Protein Data Bank". RCSB Protein Data Bank (PDB). Archived from the original on 28 August 2008. Retrieved 12 October 2012

- Phycobilins, ems373jngredler, ncsu.edu: sites.google.com, Retrieved 10 August, 2019

4

Light Reactions in Photosynthesis

The reactions which constitute the first stage of photosynthesis where light energy is converted into chemical energy are known as light-dependent reactions. The chapter closely examines the major types of light reactions in photosynthesis such as cyclic photophosphorylation and non-cyclic photophosphorylation.

LIGHT-DEPENDENT REACTIONS

The light-dependent reactions use light energy to make two molecules needed for the next stage of photosynthesis: the energy storage molecule ATP and the reduced electron carrier NADPH. In plants, the light reactions take place in the thylakoid membranes of organelles called chloroplasts.

Photosystems, large complexes of proteins and pigments (light-absorbing molecules) that are optimized to harvest light, play a key role in the light reactions. There are two types of photosystems: photosystem I (PSI) and photosystem II (PSII).

Both photosystems contain many pigments that help collect light energy, as well as a special pair of chlorophyll molecules found at the core (reaction center) of the photosystem. The special pair of photosystem I is called P700, while the special pair of photosystem II is called P680.

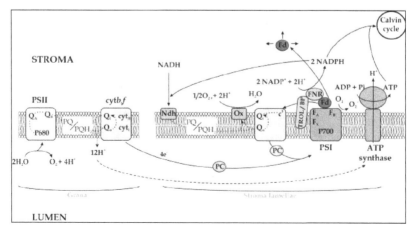

In a process called non-cyclic photophosphorylation (the "standard" form of the light-dependent reactions), electrons are removed from water and passed through PSII and PSI before ending up in NADPH. This process requires light to be absorbed twice, once in each photosystem, and it makes ATP. In fact, it's called photophosphorylation because it involves using light energy to make ATP from ADP (*phosphorylation*). Here are the basic steps:

- Light absorption in PSII: When light is absorbed by one of the many pigments in photosystem II, energy is passed inward from pigment to pigment until it reaches the reaction center. There, energy is transferred to P680, boosting an electron to a high energy level. The high-energy electron is passed to an acceptor molecule and replaced with an electron from water. This splitting of water releases the O_2 we breathe.

- ATP synthesis: The high-energy electron travels down an electron transport chain, losing energy as it goes. Some of the released energy drives pumping of H^+ ions from the stroma into the thylakoid interior, building a gradient. (H^+ ions from the splitting of water also add to the gradient.) As H^+ ions flow down their gradient and into the stroma, they pass through ATP synthase, driving ATP production in a process known as chemiosmosis.

- Light absorption in PSI: The electron arrives at photosystem I and joins the P700 special pair of chlorophylls in the reaction center. When light energy is absorbed by pigments and passed inward to the reaction center, the electron in P700 is boosted to a very high energy level and transferred to an acceptor molecule. The special pair's missing electron is replaced by a new electron from PSII (arriving via the electron transport chain).

- NADPH formation: The high-energy electron travels down a short second leg of the electron transport chain. At the end of the chain, the electron is passed to $NADP^+$ (along with a second electron from the same pathway) to make NADPH.

The net effect of these steps is to convert light energy into chemical energy in the form of ATP and NADPH. The ATP and NADPH from the light-dependent reactions are used to make sugars in the next stage of photosynthesis, the Calvin cycle. In another form of the light reactions, called cyclic photophosphorylation, electrons follow a different, circular path and only ATP (no NADPH) is produced.

It's important to realize that the electron transfers of the light-dependent reactions are driven by, and indeed made possible by, the absorption of energy from light. In other words, the transfers of electrons from PSII to PSI, and from PSI to NADPH, are only energetically "downhill" (energy-releasing, and thus spontaneous) because electrons in P680 and P700 are boosted to very high energy levels by absorption of energy from light.

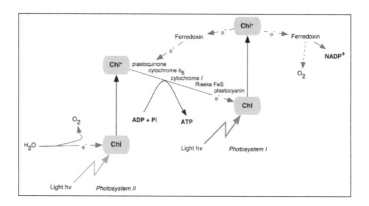

Photosystem

Photosynthetic pigments, such as chlorophyll a, chlorophyll b, and carotenoids, are light-harvesting molecules found in the thylakoid membranes of chloroplasts. As mentioned above, pigments are organized along with proteins into complexes called photosystems. Each photosystem has light-harvesting complexes that contain proteins, 300-400 chlorophylls, and other pigments. When a pigment absorbs a photon, it is raised to an excited state, meaning that one of its electrons is boosted to a higher-energy orbital.

Most of the pigments in a photosystem act as an energy funnel, passing energy inward to a main reaction center. When one of these pigments is excited by light, it transfers energy to a neighboring pigment through direct electromagnetic interactions in a process called resonance energy transfer. The neighbor pigment, in turn, can transfer energy to one of its own neighbors, with the process repeating multiple times. In these transfers, the receiving molecule cannot require more energy for excitation than the donor, but may require less energy (i.e., may absorb light of a longer wavelength).

Collectively, the pigment molecules collect energy and transfer it towards a central part of the photosystem called the reaction center.

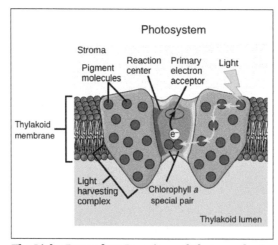

The Light-Dependent Reactions of Photosynthesis.

The reaction center of a photosystem contains a unique pair of chlorophyll *a* molecules, often called special pair. Once energy reaches the special pair, it will no longer be passed on to other pigments through resonance energy transfer. Instead, the special pair can actually lose an electron when excited, passing it to another molecule in the complex called the primary electron acceptor. With this transfer, the electron will begin its journey through an electron transport chain.

Photosystem I vs. Photosystem II

There are two types of photosystems in the light-dependent reactions, photosystem II (PSII) and photosystem I (PSI). PSII comes first in the path of electron flow, but it is named as second because it was discovered after PSI.

Here are some of the key differences between the photosystems:

*   Special pairs: The chlorophyll a special pairs of the two photosystems absorb different wavelengths of light. The PSII special pair absorbs best at 680 nm, while the PSI special absorbs best at 700 nm. Because of this, the special pairs are called P680 and P700, respectively.

*   Primary acceptor: The special pair of each photosystem passes electrons to a different primary acceptor. The primary electron acceptor of PSII is pheophytin, an organic molecule that resembles chlorophyll, while the primary electron acceptor of PSI is a chlorophyll called Ao.

*   Source of electrons: Once an electron is lost, each photosystem is replenished by electrons from a different source. The PSII reaction center gets electrons from water, while the PSI reaction center is replenished by electrons that flow down an electron transport chain from PSII.

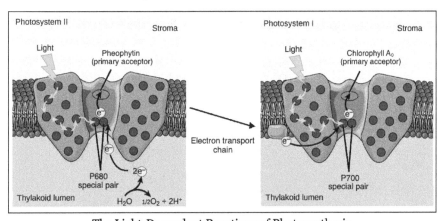

The Light-Dependent Reactions of Photosynthesis.

During the light-dependent reactions, an electron that's excited in PSII is passed down an electron transport chain to PSI (losing energy along the way). In PSI, the electron is excited again and passed down the second leg of the electron transport chain to a final electron acceptor. Let's trace the path of electrons in more detail, starting when they're excited by light energy in PSII.

Photosystem II

When the P680 special pair of photosystem II absorbs energy, it enters an excited (high-energy) state. Excited P680 is a good electron donor and can transfer its excited electron to the primary electron acceptor, pheophytin. The electron will be passed on through the first leg of the photosynthetic electron transport chain in a series of redox, or electron transfer, reactions.

After the special pair gives up its electron, it has a positive charge and needs a new electron. This electron is provided through the splitting of water molecules, a process carried out by a portion of PSII called the manganese center. The positively charged P680 can pull electrons off of water (which doesn't give them up easily) because it's extremely "electron-hungry."

When the manganese center splits water molecules, it binds two at once, extracting four electrons, releasing four H^+ ions, and producing a molecule of O_2. About 10 percent of the oxygen is used by mitochondria in the leaf to support oxidative phosphorylation. The remainder escapes to the atmosphere where it is used by aerobic organisms to support respiration.

Electron Transport Chains and Photosystem I

When an electron leaves PSII, it is transferred first to a small organic molecule (plastoquinone, Pq), then to a cytochrome complex (Cyt), and finally to a copper-containing protein called plastocyanin (Pc). As the electron moves through this electron transport chain, it goes from a higher to a lower energy level, releasing energy. Some of the energy is used to pump protons (H^+) from the stroma (outside of the thylakoid) into the thylakoid interior.

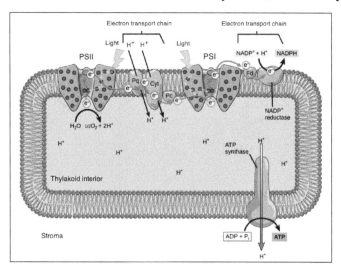

This transfer of H⁺, along with the release of H⁺ from the splitting of water, forms a proton gradient that will be used to make ATP.

Once an electron has gone down the first leg of the electron transport chain, it arrives at PSI, where it joins the chlorophyll *a* special pair called P700. Because electrons have lost energy prior to their arrival at PSI, they must be re-energized through absorption of another photon.

Excited P700 is a very good electron donor, and it sends its electron down a short electron transport chain. In this series of reactions, the electron is first passed to a protein called ferredoxin (Fd), then transferred to an enzyme called NADP⁺ reductase NADP⁺ reductase transfers electrons to the electron carrier NADP⁺ to make NADPH. NADPH will travel to the Calvin cycle, where its electrons are used to build sugars from carbon dioxide.

The other ingredient needed by the Calvin cycle is ATP, and this too is provided by the light reactions. As we saw above, H⁺ ions build inside the thylakoid interior and make a concentration gradient. Protons "want" to diffuse back down the gradient and into the stroma, and their only route of passage is through the enzyme ATP synthase. ATP synthase harnesses the flow of protons to make ATP from ADP and phosphate (Pᵢ). This process of making ATP using energy stored in a chemical gradient is called chemiosmosis.

Some Electrons Flow Cyclically

The pathway above is sometimes called linear photophosphorylation. That's because electrons travel in a line from water through PSII and PSI to NADPH. (*Photophosphorylation* = light-driven synthesis of ATP.)

In some cases, electrons break this pattern and instead loop back to the first part of the electron transport chain, repeatedly cycling through PSI instead of ending up in NADPH. This is called cyclic photophosphorylation.

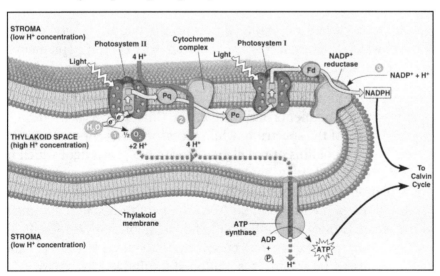

After leaving PSI, cyclically flowing electrons travel back to the cytochrome complex (Cyt) or plastoquinone (Pq) in the first leg of the electron transport chain. The electrons then flow down the chain to PSI as usual, driving proton pumping and the production of ATP. The cyclic pathway does not make NADPH, since electrons are routed away from $NADP^+$ reductase.

Why does the cyclic pathway exist? At least in some cases, chloroplasts seem to switch from linear to cyclic electron flow when the ratio of NADPH to $NADP^+$ is too high (when too little $NADP^+$ is available to accept electrons). In addition, cyclic electron flow may be common in photosynthetic cell types with especially high ATP needs (such as the sugar-synthesizing bundle-sheath cells of plants that carry out C_4 photosynthesis). Finally, cyclic electron flow may play a photoprotective role, preventing excess light from damaging photosystem proteins and promoting repair of light-induced damage.

Photolysis in Photosynthesis

Photolysis is part of the light-dependent reactions of photosynthesis. The general reaction of photosynthetic photolysis can be given as:

$$H_2A + 2 \text{ photons (light)} \rightarrow 2\ e^- + 2\ H^+ + A$$

The chemical nature of "A" depends on the type of organism. In purple sulfur bacteria, hydrogen sulfide (H_2S) is oxidized to sulfur (S). In oxygenic photosynthesis, water (H_2O) serves as a substrate for photolysis resulting in the generation of diatomic oxygen (O_2). This is the process which returns oxygen to Earth's atmosphere. Photolysis of water occurs in the thylakoids of cyanobacteria and the chloroplasts of green algae and plants.

Energy Transfer Models

The conventional, semi-classical, model describes the photosynthetic energy transfer process as one in which excitation energy hops from light-capturing pigment molecules to reaction center molecules step-by-step down the molecular energy ladder.

The effectiveness of photons of different wavelengths depends on the absorption spectra of the photosynthetic pigments in the organism. Chlorophylls absorb light in the violet-blue and red parts of the spectrum, while accessory pigments capture other wavelengths as well. The phycobilins of red algae absorb blue-green light which penetrates deeper into water than red light, enabling them to photosynthesize in deep waters. Each absorbed photon causes the formation of an exciton (an electron excited to a higher energy state) in the pigment molecule. The energy of the exciton is transferred to a chlorophyll molecule (P680, where P stands for pigment and 680 for its absorption maximum at 680 nm) in the reaction center of photosystem II via resonance energy transfer. P680 can also directly absorb a photon at a suitable wavelength.

Photolysis during photosynthesis occurs in a series of light-driven oxidation events. The energized electron (exciton) of P680 is captured by a primary electron acceptor of the photosynthetic electron transfer chain and thus exits photosystem II. In order to repeat the reaction, the electron in the reaction center needs to be replenished. This occurs by oxidation of water in the case of oxygenic photosynthesis. The electron-deficient reaction center of photosystem II (P680) is the strongest biological oxidizing agent yet discovered, which allows it to break apart molecules as stable as water.

The water-splitting reaction is catalyzed by the oxygen evolving complex of photosystem II. This protein-bound inorganic complex contains four manganese ions, plus calcium and chloride ions as cofactors. Two water molecules are complexed by the manganese cluster, which then undergoes a series of four electron removals (oxidations) to replenish the reaction center of photosystem II. At the end of this cycle, free oxygen (O_2) is generated and the hydrogen of the water molecules has been converted to four protons released into the thylakoid lumen.

These protons, as well as additional protons pumped across the thylakoid membrane coupled with the electron transfer chain, form a proton gradient across the membrane that drives photophosphorylation and thus the generation of chemical energy in the form of adenosine triphosphate (ATP). The electrons reach the P700 reaction center of photosystem I where they are energized again by light. They are passed down another electron transfer chain and finally combine with the coenzyme $NADP^+$ and protons outside the thylakoids to form NADPH. Thus, the net oxidation reaction of water photolysis can be written as:

$$2\ H_2O + 2\ NADP^+ + 8\ \text{photons (light)} \rightarrow 2\ NADPH + 2\ H^+ + O_2$$

The free energy change (ΔG) for this reaction is 102 kilocalories per mole. Since the energy of light at 700 nm is about 40 kilocalories per mole of photons, approximately 320 kilocalories of light energy are available for the reaction. Therefore, approximately one-third of the available light energy is captured as NADPH during photolysis and electron transfer. An equal amount of ATP is generated by the resulting proton gradient. Oxygen as a byproduct is of no further use to the reaction and thus released into the atmosphere.

PHOTOPHOSPHORYLATION

In the process of photosynthesis, the phosphorylation of ADP to form ATP using the energy of sunlight is called photophosphorylation. Only two sources of energy are available to living organisms: sunlight and reduction-oxidation (redox) reactions. All organisms produce ATP, which is the universal energy currency of life. In photosynthesis this commonly involves photolysis, or photodissociation, of water and a continuous unidirectional flow of electrons from water to photosystem II.

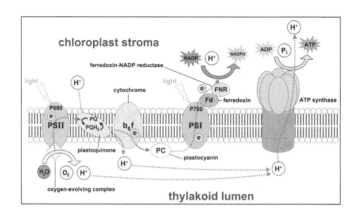

The scientist Charles Barnes first used the word 'photosynthesis' in 1893. This word is taken from two Greek words, *photos* which means light and *synthesis* which in chemistry means making a substance by combining simpler substances. So, in the presence of light, synthesis of food is called 'photosynthesis'. Noncyclic photophosphorylation through light-dependent reactions of photosynthesis at the thylakoid membrane.

In photophosphorylation, light energy is used to create a high-energy electron donor and a lower-energy electron acceptor. Electrons then move spontaneously from donor to acceptor through an electron transport chain.

ATP and Reactions

ATP is made by an enzyme called ATP synthase. Both the structure of this enzyme and its underlying gene are remarkably similar in all known forms of life. The Calvin cycle is one of the most important part of photosynthesis.

ATP synthase is powered by a transmembrane electrochemical potential gradient, usually in the form of a proton gradient. The function of the electron transport chain is to produce this gradient. In all living organisms, a series of redox reactions is used to produce a transmembrane electrochemical potential gradient, or a so-called proton motive force (pmf).

Redox reactions are chemical reactions in which electrons are transferred from a donor molecule to an acceptor molecule. The underlying force driving these reactions is the Gibbs free energy of the reactants and products. The Gibbs free energy is the energy available ("free") to do work. Any reaction that decreases the overall Gibbs free energy of a system will proceed spontaneously (given that the system is isobaric and also adiabatic), although the reaction may proceed slowly if it is kinetically inhibited.

The transfer of electrons from a high-energy molecule (the donor) to a lower-energy molecule (the acceptor) can be *spatially* separated into a series of intermediate redox reactions. This is an electron transport chain.

The fact that a reaction is thermodynamically possible does not mean that it will actually occur. A mixture of hydrogen gas and oxygen gas does not spontaneously ignite. It is necessary either to supply an activation energy or to lower the intrinsic activation energy of the system, in order to make most biochemical reactions proceed at a useful rate. Living systems use complex macromolecular structures to lower the activation energies of biochemical reactions.

It is possible to couple a thermodynamically favorable reaction (a transition from a high-energy state to a lower-energy state) to a thermodynamically unfavorable reaction (such as a separation of charges, or the creation of an osmotic gradient), in such a way that the overall free energy of the system decreases (making it thermodynamically possible), while useful work is done at the same time. The principle that biological macromolecules catalyze a thermodynamically unfavorable reaction *if and only if* a thermodynamically favorable reaction occurs simultaneously, underlies all known forms of life.

Electron transport chains (most known as ETC) produce energy in the form of a transmembrane electrochemical potential gradient. This energy is used to do useful work. The gradient can be used to transport molecules across membranes. It can be used to do mechanical work, such as rotating bacterial flagella. It can be used to produce ATP and NADPH, high-energy molecules that are necessary for growth.

Cyclic Photophosphorylation

This form of photophosphorylation occurs on the stroma lamella or fret channels. In cyclic photophosphorylation, the high energy electron released from P700 to ps1 flow down in a cyclic pathway. In cyclic electron flow, the electron begins in a pigment complex called photosystem I, passes from the primary acceptor to ferredoxin and then to plastoquinone, then to cytochrome b_6f (a similar complex to that found in mitochondria), and then to plastocyanin before returning to Photosystem-1. This transport chain produces a proton-motive force, pumping H^+ ions across the membrane; this produces a concentration gradient that can be used to power ATP synthase during chemiosmosis. This pathway is known as cyclic photophosphorylation, and it produces neither O_2 nor NADPH. Unlike non-cyclic photophosphorylation, $NADP^+$ does not accept the electrons; they are instead sent back to cytochrome b_6f complex.

In bacterial photosynthesis, a single photosystem is used, and therefore is involved in cyclic photophosphorylation. It is favored in anaerobic conditions and conditions of high irradiance and CO_2 compensation points.

Non-cyclic Photophosphorylation

The other pathway, non-cyclic photophosphorylation, is a two-stage process involving two different chlorophyll photosystems. Being a light reaction, non-cyclic photophosphorylation occurs in the thylakoid membrane. First, a water molecule is broken down into $2H^+ + 1/2\ O_2 + 2e^-$ by a process called photolysis (or *light-splitting*). The two electrons from the water molecule are kept in photosystem II, while the $2H^+$ and $1/2O_2$ are left out for further use. Then a photon is absorbed by chlorophyll pigments surrounding the reaction core center of the photosystem. The light excites the electrons of each pigment, causing a chain reaction that eventually transfers energy to the core of photosystem II, exciting the two electrons that are transferred to the primary electron acceptor, pheophytin. The deficit of electrons is replenished by taking electrons from another molecule of water. The electrons transfer from pheophytin to plastoquinone, which takes the $2e^-$ from Pheophytin, and two H^+ Ions from the stroma and forms PQH_2, which later is broken into PQ, the $2e^-$ is released to Cytochrome b_6f complex and the two H^+ ions are released into thylakoid lumen. The electrons then pass through the Cyt b_6 and Cyt f. Then they are passed to plastocyanin, providing the energy for hydrogen ions (H^+) to be pumped into the thylakoid space. This creates a gradient, making H^+ ions flow back into the stroma of the chloroplast, providing the energy for the regeneration of ATP.

The photosystem II complex replaced its lost electrons from an external source; however, the two other electrons are not returned to photosystem II as they would in the analogous cyclic pathway. Instead, the still-excited electrons are transferred to a photosystem I complex, which boosts their energy level to a higher level using a second solar photon. The highly excited electrons are transferred to the acceptor molecule, but this time are passed on to an enzyme called Ferredoxin-NADP$^+$ reductase which uses them to catalyse the reaction:

$$NADP^+ + 2H^+ + 2e^- \rightarrow NADPH + H^+$$

This consumes the H^+ ions produced by the splitting of water, leading to a net production of $1/2O_2$, ATP, and $NADPH+H^+$ with the consumption of solar photons and water.

The concentration of NADPH in the chloroplast may help regulate which pathway electrons take through the light reactions. When the chloroplast runs low on ATP for the Calvin cycle, NADPH will accumulate and the plant may shift from noncyclic to cyclic electron flow.

PHOTOSYNTHETIC REACTION CENTRE

A photosynthetic reaction center is a complex of several proteins, pigments and other co-factors that together execute the primary energy conversion reactions of photosynthesis. Molecular excitations, either originating directly from sunlight or transferred as excitation energy via light-harvesting antenna systems, give rise to electron transfer reactions along the path of a series of protein-bound co-factors. These co-factors are light-absorbing molecules (also named chromophores or pigments) such as chlorophyll and phaeophytin, as well as quinones. The energy of the photon is used to excite an electron of a pigment. The free energy created is then used to reduce a chain of nearby electron acceptors, which have subsequently higher redox-potentials. These electron transfer steps are the initial phase of a series of energy conversion reactions, ultimately resulting in the conversion of the energy of photons to the storage of that energy by the production of chemical bonds.

Electron micrograph of 2D crystals of the LH1-Reaction center photosynthetic unit.

Transforming Light Energy into Charge Separation

Reaction centers are present in all green plants, algae, and many bacteria. Although these species are separated by billions of years of evolution, the reaction centers are homologous for all photosynthetic species. In contrast, a large variety in light-harvesting complexes exist between the photosynthetic species. Green plants and algae have two different types of reaction centers that are part of larger supercomplexes known as photosystem I P700 and photosystem II P680. The structures of these supercomplexes are large, involving multiple light-harvesting complexes. The reaction center found in *Rhodopseudomonas* bacteria is currently best understood, since it was the first reaction center of known structure and has fewer polypeptide chains than the examples in green plants.

A reaction center is laid out in such a way that it captures the energy of a photon using pigment molecules and turns it into a usable form. Once the light energy has been absorbed directly by the pigment molecules, or passed to them by resonance transfer from a surrounding light-harvesting complex, they release two electrons into an electron transport chain.

Light is made up of small bundles of energy called photons. If a photon with the right amount of energy hits an electron, it will raise the electron to a higher energy level. Electrons are most stable at their lowest energy level, what is also called its ground state. In this state, the electron is in the orbit that has the least amount of energy.

Electrons in higher energy levels can return to ground state in a manner analogous to a ball falling down a staircase. In doing so, the electrons release energy. This is the process that is exploited by a photosynthetic reaction center.

When an electron rises to a higher energy level, there is a corresponding decrease in the reduction potential of the molecule in which the electron resides occurs. This means that the molecule has a greater tendency to donate electrons, which is key to the conversion of light energy to chemical energy. In green plants, the electron transport chain has many electron acceptors including phaeophytin, quinone, plastoquinone, cytochrome bf, and ferredoxin, which result finally in the reduced molecule NADPH and the storage of energy. The passage of the electron through the electron transport chain also results in the pumping of protons (hydrogen ions) from the chloroplast's stroma and into the lumen, resulting in a proton gradient across the thylakoid membrane that can be used to synthesise ATP using the ATP synthase molecule. Both the ATP and NADPH are used in the Calvin cycle to fix carbon dioxide into triose sugars.

In bacteria

Classification

Two classes of reaction centres are recognized. Type I, found in green-sulfur bacteria, Heliobacteria, and plant/cyanobacterial PS-I, use iron sulfur clusters as electron acceptors. Type II, found in chloroflexus, purple bacteria, and plant/cyanobacterial PS-II, use quinones. Not only do all members inside each class share common ancestry, but the two classes also, by means of common structure, appear related.

Structure

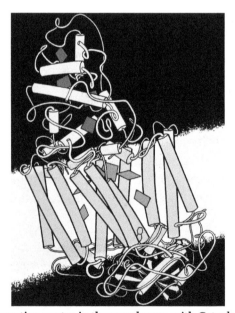

Schematic of reaction center in the membrane, with Cytochrome C at top.

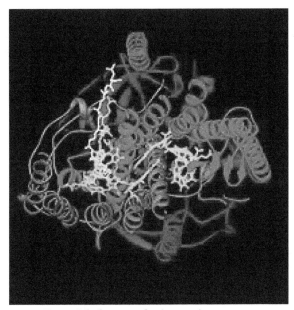

Bacterial photosynthetic reaction center.

The bacterial photosynthetic reaction center has been an important model to understand the structure and chemistry of the biological process of capturing light energy. In the 1960s, Roderick Clayton was the first to purify the reaction center complex from purple bacteria. However, the first crystal structure was determined in 1984 by Hartmut Michel, Johann Deisenhofer and Robert Huber for which they shared the Nobel Prize in 1988. This was also significant for being the first 3D crytal structure of any membrane protein complex.

Four different subunits were found to be important for the function of the photosynthetic reaction center. The L and M subunits, shown in blue and purple in the image of the structure, both span the lipid bilayer of the plasma membrane. They are structurally similar to one another, both having 5 transmembrane alpha helices. Four bacteriochlorophyll b (BChl-b) molecules, two bacteriophaeophytin b molecules (BPh) molecules, two quinones (Q_A and Q_B), and a ferrous ion are associated with the L and M subunits. The H subunit, shown in gold, lies on the cytoplasmic side of the plasma membrane. A cytochrome subunit, contains four c-type haems and is located on the periplasmic surface (outer) of the membrane. The latter sub-unit is not a general structural motif in photosynthetic bacteria. The L and M subunits bind the functional and light-interacting cofactors.

Reaction centers from different bacterial species may contain slightly altered bacterio-chlorophyll and bacterio-phaeophytin chromophores as functional co-factors. These alterations cause shifts in the colour of light that can be absorbed, thus creating specific niches for photosynthesis. The reaction center contains two pigments that serve to collect and transfer the energy from photon absorption: BChl and Bph. BChl roughly resembles the chlorophyll molecule found in green plants, but, due to minor structural differences, its peak absorption wavelength is shifted into the infrared, with wavelengths as long as 1000 nm. Bph has the same structure as BChl, but the central magnesium ion is replaced by two protons. This alteration causes both an absorbance maximum shift and a lowered redox-potential.

Mechanism

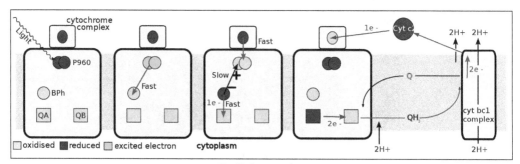

The light reaction

The process starts when light is absorbed by two BChl molecules (a dimer) that lie near the periplasmic side of the membrane. This pair of chlorophyll molecules, often called the "special pair", absorbs photons between 870 nm and 960 nm, depending on the species and, thus, is called P870 (for *Rhodobacter sphaeroides*) or P960 (for *Blastochloris viridis*), with *P* standing for "pigment"). Once P absorbs a photon, it ejects an electron, which is transferred through another molecule of Bchl to the BPh in the L subunit. This initial charge separation yields a positive charge on P and a negative charge on the BPh. This process takes place in 10 picoseconds (10^{-11} seconds).

The charges on the specialpair $^+$ and the BPh$^-$ could undergo charge recombination in this state. This would waste the high-energy electron and convert the absorbed light energy into heat. Several factors of the reaction center structure serve to prevent this. First, the transfer of an electron from BPh$^-$ to P960$^+$ is relatively slow compared to two other redox reactions in the reaction center. The faster reactions involve the transfer of an electron from BPh$^-$ (BPh$^-$ is oxidised to BPh) to the electron acceptor quinone (Q_A), and the transfer of an electron to P960$^+$ (P960$^+$ is reduced to P960) from a heme in the cytochrome subunit above the reaction center.

The high-energy electron that resides on the tightly bound quinone molecule Q_A is transferred to an exchangeable quinone molecule Q_B. This molecule is loosely associated with the protein and is fairly easy to detach. Two of the high-energy electrons are required to fully reduce Q_B to QH_2, taking up two protons from the cytoplasm in the process. The reduced quinone QH_2 diffuses through the membrane to another protein complex (cytochrome bc$_1$-complex) where it is oxidised. In the process the reducing power of the QH_2 is used to pump protons across the membrane to the periplasmic space. The electrons from the cytochrome bc$_1$-complex are then transferred through a soluble cytochrome c intermediate, called cytochrome c_2, in the periplasm to the cytochrome subunit. Thus, the flow of electrons in this system is cyclical.

In Cyanobacteria and Plants

Cyanobacteria, the precursor to chloroplasts found in green plants, have both photo-systems with both types of reaction centers. Combining the two systems allows for pro-ducing oxygen.

Oxygenic Photosynthesis

In 1772, the chemist Joseph Priestley carried out a series of experiments relating to the gases involved in respiration and combustion. In his first experiment, he lit a candle and placed it under an upturned jar. After a short period of time, the candle burned out. He carried out a similar experiment with a mouse in the confined space of the burning candle. He found that the mouse died a short time after the candle had been extinguished. However, he could revivify the foul air by placing green plants in the area and exposing them to light. Priestley's observations were some of the first experiments that demonstrated the activity of a photosynthetic reaction center.

In 1779, Jan Ingenhousz carried out more than 500 experiments spread out over 4 months in an attempt to understand what was really going on. Ingenhousz took green plants and immersed them in water inside a transparent tank. He observed many bubbles rising from the surface of the leaves whenever the plants were exposed to light. Ingenhousz collected the gas that was given off by the plants and performed several different tests in attempt to determine what the gas was. The test that finally revealed the identity of the gas was placing a smouldering taper into the gas sample and having it relight. This test proved it was oxygen, or, as Joseph Priestley had called it, 'de-phlogisticated air'.

In 1932, Professor Robert Emerson and an undergraduate student, William Arnold, used a repetitive flash technique to precisely measure small quantities of oxygen evolved by chlorophyll in the algae *Chlorella*. Their experiment proved the existence of a photosynthetic unit. Gaffron and Wohl later interpreted the experiment and realized that the light absorbed by the photosynthetic unit was transferred. This reaction occurs at the reaction center of photosystem II and takes place in cyanobacteria, algae and green plants.

Photosystem II

Photosystem II is the photosystem that generates the two electrons that will eventually reduce $NADP^+$ in ferredoxin-NADP-reductase. Photosystem II is present on the thyla-koid membranes inside chloroplasts, the site of photosynthesis in green plants. The structure of photosystem II is remarkably similar to the bacterial reaction center, and it is theorized that they share a common ancestor.

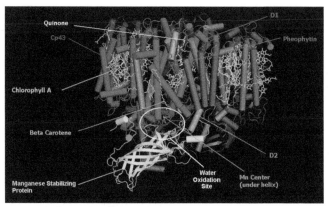

Cyanobacteria photosystem II, Monomer, PDB 2AXT.

The core of photosystem II consists of two subunits referred to as D1 and D2. These two subunits are similar to the L and M subunits present in the bacterial reaction center. Photosystem II differs from the bacterial reaction center in that it has many additional subunits that bind additional chlorophylls to increase efficiency. The overall reaction catalysed by photosystem II is:

$$2Q + 2H_2O + hv \rightarrow O_2 + 2QH_2$$

Q represents plastoquinone, the oxidized form of Q. QH_2 represents plastoquinol, the reduced form of Q. This process of reducing quinone is comparable to that which takes place in the bacterial reaction center. Photosystem II obtains electrons by oxidizing water in a process called photolysis. Molecular oxygen is a byproduct of this process, and it is this reaction that supplies the atmosphere with oxygen. The fact that the oxygen from green plants originated from water was first deduced by the Canadian-born American biochemist Martin David Kamen. He used a natural, stable isotope of oxygen, O_{18} to trace the path of the oxygen, from water to gaseous molecular oxygen. This reaction is catalysed by a reactive center in photosystem II containing four manganese ions.

The reaction begins with the excitation of a pair of chlorophyll molecules similar to those in the bacterial reaction center. Due to the presence of chlorophyll a, as opposed to bacteriochlorophyll, photosystem II absorbs light at a shorter wavelength. The pair of chlorophyll molecules at the reaction center are often referred to as P680. When the photon has been absorbed, the resulting high-energy electron is transferred to a nearby phaeophytin molecule. This is above and to the right of the pair on the diagram and is coloured grey. The electron travels from the phaeophytin molecule through two plastoquinone molecules, the first tightly bound, the second loosely bound. The tightly bound molecule is shown above the phaeophytin molecule and is coloured red. The loosely bound molecule is to the left of this and is also coloured red. This flow of electrons is similar to that of the bacterial reaction center. Two electrons are required to fully reduce the loosely bound plastoquinone molecule to QH_2 as well as the uptake of two protons.

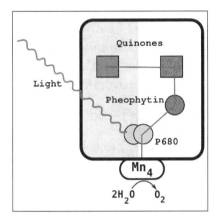

The difference between photosystem II and the bacterial reaction center is the source of the electron that neutralizes the pair of chlorophyll a molecules. In the bacterial reaction center, the electron is obtained from a reduced compound haem group in a cytochrome subunit or from a water-soluble cytochrome-c protein.

Once photoinduced charge separation has taken place, the P680 molecule carries a positive charge. P680 is a very strong oxidant and extracts electrons from two water molecules that are bound at the manganese center directly below the pair. This center, below and to the left of the pair in the diagram, contains four manganese ions, a calcium ion, a chloride ion, and a tyrosine residue. Manganese is efficient because it is capable of existing in four oxidation states: Mn^{2+}, Mn^{3+}, Mn^{4+} and Mn^{5+}. Manganese also forms strong bonds with oxygen-containing molecules such as water.

Every time the P680 absorbs a photon, it emits an electron, gaining a positive charge. This charge is neutralized by the extraction of an electron from the manganese center, which sits directly below it. The process of oxidizing two molecules of water requires four electrons. The water molecules that are oxidized in the manganese center are the source of the electrons that reduce the two molecules of Q to QH_2. To date, this water-splitting catalytic center cannot be reproduced by any man-made catalyst.

Photosystem I

After the electron has left photosystem II it is transferred to a cytochrome b6f complex and then to plastocyanin, a blue copper protein and electron carrier. The plastocyanin complex carries the electron that will neutralize the pair in the next reaction center, photosystem I.

As with photosystem II and the bacterial reaction center, a pair of chlorophyll a molecules initiates photoinduced charge separation. This pair is referred to as P700. 700 Is a reference to the wavelength at which the chlorophyll molecules absorb light maximally. The P700 lies in the center of the protein. Once photoinduced charge separation has been initiated, the electron travels down a pathway through a chlorophyll α molecule situated directly above the P700, through a quinone molecule situated directly above that, through three 4Fe-4S clusters, and finally to an interchangeable ferredoxin complex. Ferredoxin is a soluble protein containing a 2Fe-2S cluster coordinated by four cysteine residues. The positive charge left on the P700 is neutralized by the transfer of an electron from plastocyanin. Thus the overall reaction catalysed by photosystem I is:

$$Pc(Cu^+) + Fd[ox] + h\nu \rightarrow Pc(Cu^{2+}) + Fd[red]$$

The cooperation between photosystems I and II creates an electron flow from H_2O to $NADP^+$. This pathway is called the 'Z-scheme' because the redox diagram from P680 to P700 resembles the letter z.

LIGHT-INDEPENDENT REACTIONS

Light-independent reactions take place in plant chloroplasts. In this process, sugars are made from carbon dioxide. The process, known as the Calvin cycle, uses products of the light-dependent reactions (ATP and NADPH) and various enzymes. Therefore, the light-independent reaction cannot happen without the light-dependent reaction. Sugars made in the light-independent reactions are then moved around the plant by translocation.

Cactus: The harsh conditions of the desert have led plants like these cacti to evolve variations of the light-independent reactions of photosynthesis. These variations increase the efficiency of water usage, helping to conserve water and energy.

Some plants have evolved mechanisms to increase the CO_2 concentration in their leaves under hot and dry conditions.

Photosynthesis in desert plants has evolved adaptations that conserve water. In harsh, dry heat, every drop of water must be used to survive. Because stomata must open to allow for the uptake of CO_2, water escapes from the leaf during active photosynthesis. Desert plants have evolved processes to conserve water and deal with harsh conditions. A more efficient use of CO_2 allows plants to adapt to living with less water.

Some plants such as cacti can prepare materials for photosynthesis during the night by a temporary carbon fixation and storage process, because opening the stomata at this time conserves water due to cooler temperatures. In addition, cacti have evolved the ability to carry out low levels of photosynthesis without opening stomata at all, a mechanism for surviving extremely dry periods.

CAM Photosynthesis

Xerophytes, such as cacti and most succulents, also use phosphoenolpyruvate (PEP) carboxylase to capture carbon dioxide in a process called crassulacean acid metabolism (CAM). In contrast to C_4 metabolism, which *physically* separates the CO_2 fixation to PEP from the Calvin cycle, CAM *temporally* separates these two processes.

CAM plants have a different leaf anatomy from C_3 plants, and fix the CO_2 at night, when their stomata are open. CAM plants store the CO_2 mostly in the form of malic acid via carboxylation of phosphoenolpyruvate to oxaloacetate, which is then reduced to malate. Decarboxylation of malate during the day releases CO_2 inside the leaves, thus allowing carbon fixation to 3-phosphoglycerate by RuBisCO. Sixteen thousand species of plants use CAM.

C_4 Carbon Fixation

The C_4 pathway bears resemblance to CAM; both act to concentrate CO_2 around RuBisCO, thereby increasing its efficiency. CAM concentrates it temporally, providing CO_2 during the day and not at night, when respiration is the dominant reaction.

C_4 plants, in contrast, concentrate CO_2 spatially, with a RuBisCO reaction centre in a "bundle sheath cell" that is inundated with CO_2. Due to the inactivity required by the CAM mechanism, C_4 carbon fixation has a greater efficiency in terms of PGA synthesis.

C_4 plants can produce more sugar than C_3 plants in conditions of high light and temperature. Many important crop plants are C_4 plants, including maize, sorghum, sugarcane, and millet. Plants that do not use PEP-carboxylase in carbon fixation are called C_3 plants because the primary carboxylation reaction, catalyzed by RuBisCO, produces the three-carbon 3-phosphoglyceric acids directly in the Calvin-Benson cycle. Over 90% of plants use C_3 carbon fixation, compared to 3% that use C_4 carbon fixation; however, the evolution of C_4 in over 60 plant lineages makes it a striking example of convergent evolution.

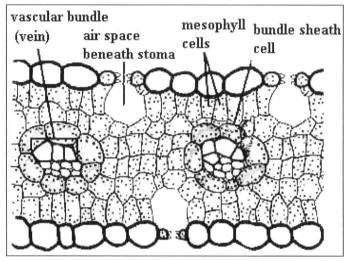

Cross section of a C_4 plant.

The Calvin Cycle

In plants, carbon dioxide (CO_2) enters the leaves through stomata, where it diffuses over short distances through intercellular spaces until it reaches the mesophyll cells. Once in the mesophyll cells, CO_2 diffuses into the stroma of the chloroplast, the site of light-independent reactions of photosynthesis. These reactions actually have several names associated with them. Other names for light-independent reactions include the Calvin cycle, the Calvin-Benson cycle, and dark reactions. The most outdated name is dark reactions, which can be misleading because it implies incorrectly that the reaction only occurs at night or is independent of light, which is why most scientists and instructors no longer use it.

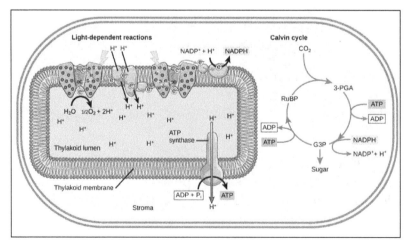

Light Reactions: Light-dependent reactions harness energy from the sun to produce chemical bonds, ATP, and NADPH. These energy-carrying molecules are made in the stroma where the Calvin cycle takes place. The Calvin cycle is not totally independent of light since it relies on ATP and NADH, which are products of the light-dependent reactions.

The light-independent reactions of the Calvin cycle can be organized into three basic stages: fixation, reduction, and regeneration.

Stage 1: Fixation

In the stroma, in addition to CO_2, two other components are present to initiate the light-independent reactions: an enzyme called ribulose bisphosphate carboxylase (RuBisCO) and three molecules of ribulose bisphosphate (RuBP). RuBP has five atoms of carbon, flanked by two phosphates. RuBisCO catalyzes a reaction between CO_2 and RuBP. For each CO_2 molecule that reacts with one RuBP, two molecules of 3-phosphoglyceric acid (3-PGA) form. 3-PGA has three carbons and one phosphate. Each turn of the cycle involves only one RuBP and one carbon dioxide and forms two molecules of 3-PGA. The number of carbon atoms remains the same, as the atoms move to form new bonds during the reactions (3 atoms from $3CO_2$ + 15 atoms from 3RuBP = 18 atoms in 3 atoms of 3-PGA). This process is called carbon fixation because CO_2 is "fixed" from an inorganic form into organic molecules.

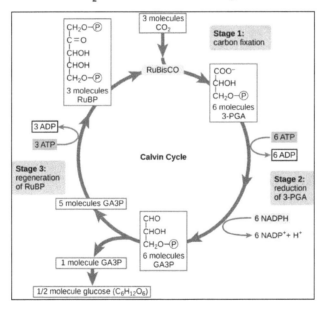

The Calvin Cycle: The Calvin cycle has three stages. In stage 1, the enzyme RuBisCO incorporates carbon dioxide into an organic molecule, 3-PGA. In stage 2, the organic molecule is reduced using electrons supplied by NADPH. In stage 3, RuBP, the molecule that starts the cycle, is regenerated so that the cycle can continue. Only one carbon dioxide molecule is incorporated at a time, so the cycle must be completed three times to produce a single three-carbon GA3P molecule, and six times to produce a six-carbon glucose molecule.

Stage 2: Reduction

ATP and NADPH are used to convert the six molecules of 3-PGA into six molecules of a chemical called glyceraldehyde 3-phosphate (G3P). This is a reduction reaction because it involves the gain of electrons by 3-PGA. Recall that a reduction is the gain of an electron by an atom or molecule. Six molecules of both ATP and NADPH are used. For ATP, energy is released with the loss of the terminal phosphate atom, converting it to ADP; for NADPH, both energy and a hydrogen atom are lost, converting it into NADP$^+$. Both of these molecules return to the nearby light-dependent reactions to be reused and reenergized.

Stage 3: Regeneration

At this point, only one of the G3P molecules leaves the Calvin cycle and is sent to the cytoplasm to contribute to the formation of other compounds needed by the plant. Because the G3P exported from the chloroplast has three carbon atoms, it takes three "turns" of the Calvin cycle to fix enough net carbon to export one G3P. But each turn makes two G3Ps, thus three turns make six G3Ps. One is exported while the remaining five G3P molecules remain in the cycle and are used to regenerate RuBP, which enables the system to prepare for more CO_2 to be fixed. Three more molecules of ATP are used in these regeneration reactions.

The Carbon Cycle

Whether the organism is a bacterium, plant, or animal, all living things access energy by breaking down carbohydrate molecules. But if plants make carbohydrate molecules, why would they need to break them down, especially when it has been shown that the gas organisms release as a "waste product" (CO_2) acts as a substrate for the formation of more food in photosynthesis? Living things need energy to perform life functions. In addition, an organism can either make its own food or eat another organism; either way, the food still needs to be broken down. Finally, in the process of breaking down food, called cellular respiration, heterotrophs release needed energy and produce "waste" in the form of CO_2 gas.

Photosynthesis and Aerobic Respiration: Photosynthesis consumes carbon dioxide and produces oxygen. Aerobic respiration consumes oxygen and produces carbon dioxide. These two processes play an important role in the carbon cycle.

In nature, there is no such thing as waste. Every single atom of matter and energy is conserved, recycling over and over, infinitely. Substances change form or move from one type of molecule to another, but their constituent atoms never disappear.

CO_2 is no more a form of waste than oxygen is wasteful to photosynthesis. Both are byproducts of reactions that move on to other reactions. Photosynthesis absorbs light energy to build carbohydrates in chloroplasts, and aerobic cellular respiration releases energy by using oxygen to metabolize carbohydrates in the cytoplasm and mitochondria. Photosynthesis consumes carbon dioxide and produces oxygen. Aerobic respiration consumes oxygen and produces carbon dioxide. Both processes use electron transport chains to capture the energy necessary to drive other reactions. These two powerhouse processes, photosynthesis and cellular respiration, function in biological, cyclical harmony to allow organisms to access life-sustaining energy that originates millions of miles away in the sun.

References

- Dolai, U. (2017). "Chemical Scheme of Water -Splitting Process during Photosynthesis by the way of Experimental Analysis". IOSR Journal of Pharmacy and Biological Sciences. 12(6):65-67. Doi: 10.9790/3008-1206026567. ISSN 2319-7676

- A light-dependent-reactions, the-light-dependent-reactions-of-photosynthesis, photosynthesis-in-plants, biology, science: khanacademy.org, Retrieved 11 January, 2019

- Gregory D. Scholes (7 January 2010), "Quantum-coherent electronic energy transfer: Did Nature think of it first?", Journal of Physical Chemistry Letters, 1 (1): 2–8, doi:10.1021/jz900062f

- Sadekar, S; Raymond, J; Blankenship, RE (November 2006). "Conservation of distantly related membrane proteins: photosynthetic reaction centers share a common structural core". Molecular Biology and Evolution. 23 (11): 2001–7. Doi:10.1093/molbev/msl079. PMID 16887904

- The-light-independent-reactions-of-photosynthesis, boundless-biology: courses.lumenlearning.com, Retrieved 12 February, 2019

- Jagannathan B, Golbeck J (2009). "Photosynthesis: microbial". In Schaechter M (ed.). Encyclopedia of Microbiology(3rd ed.). Pp. 325–341. Doi:10.1016/B978-012373944-5.00352-7. ISBN 978-0-12-373944-5

- Chapter 5 Phototrophic Bacteria

- J.A. Eisen; Nelson, KE; Paulsen, IT; Heidelberg, JF; Wu, M; Dodson, RJ; Deboy, R; Gwinn, ML; et al. (2002). "The complete genome sequence of Chlorobium tepidum TLS, a photosynthetic, anaerobic, green-sulfur bacterium". Proc. Natl. Acad. Sci. USA. 99 (14): 9509–9514. Doi:10.1073/pnas.132181499. PMC 123171. PMID 12093901

- Cyanobacteria: microscopemaster.com, Retrieved 13 April, 2019

- E., Blankenship, Robert (2002). Molecular mechanisms of photosynthesis. Oxford: Blackwell Science. ISBN 9780632043217. OCLC 49273347

- Ioanna A. Vasiliadou et al. (13 November 2018). "Biological and Bioelectrochemical Systems for Hydrogen Production and Carbon Fixation Using Purple Phototrophic Bacteria". Frontiers in Energy Research. 6. doi:10.3389/fenrg.2018.00107

5

Phototrophic Bacteria

The group of bacteria which derive energy from the sun in order to grow are known as phototrophic bacteria. A few examples of such bacteria are heliobacteria, chlorobium, chlorobium chlorochromatii, chloroflexus aurantiacus, green sulfur bacteria, rhodobacter sphaeroides, purple bacteria and cyanobacteria. This chapter has been carefully written to provide an easy understanding of these types of phototrophic bacteria.

HELIOBACTERIA

Heliobacteria are photosynthetic bacteria that uniquely employ bacteriochlorophyll (Bchl) g as the major antenna pigment and primary electron donor within a type I reaction center (RC) . Bchl g is related to chlorophyll (Chl) a but has an ethylidine functional group at the C-8^1 position and is esterified with farnesol rather than phytol. An oxidized form of Chl a, 8^1-hydroxy-Chl a, is the primary electron acceptor from the RC special pair . Unlike other anoxygenic phototrophic bacteria, heliobacteria have no Bchl-containing internal membranes or structures, such as lamellae (purple bacteria) or chlorosomes (green bacteria). In the heliobacteria, photosynthetic pigments are confined to RCs in the cytoplasmic membrane . Carotenoids are also unusual in heliobacteria in that they consist of C_{30} pigments rather than the C_{40} derivatives present in other phototrophs . The dominant carotenoid in nonalkaliphilic heliobacteria is 4,4'-diaponeurosporene.

In addition to their unique photosynthetic properties, heliobacteria can be distinguished from all other anaerobic anoxygenic phototrophs in at least three major ways. In terms of carbon metabolism, heliobacteria are obligately heterotrophic. Growth occurs either photoheterotrophically (anoxic, with light) on a limited range of organic substrates or by fermentation of pyruvate in the dark . By contrast, autotrophic growth, the hallmark of photosynthetic organisms, has not been observed with cultures of any heliobacterium species . Heliobacteria are phylogenetically unique, as they are the only phototrophic organisms that group within the bacterial phylum *Firmicutes* using 16S rRNA gene sequence analyses . Finally, heliobacteria are unique among all phototrophs in that they produce endospores, a key property of nonphototrophic *Firmicutes*, such as *Bacillus* and *Clostridium*.

A complete genome sequence analysis of *Heliobacterium modesticaldum*, the first heliobacterial genome to be sequenced, reveals an organism with a full complement of nitrogen fixation genes but only a limited capacity for carbon metabolism and no apparent mechanism for autotrophic growth. Many genes linked to endosporulation in *Bacillus subtilis* were not found in *H. modesticaldum*, which may have relevance for the ambiguous sporulation patterns observed in heliobacterial cultures.

CHLOROBIUM

Chlorobium (also known as *Chlorochromatium*) is a genus of green sulfur bacteria. They are photolithotrophic oxidizers of sulfur and most notably utilise a noncyclic electron transport chain to reduce NAD+. Photosynthesis is achieved using a Type 1 Reaction Centre using bacteriochlorophyll (BChl) *a*. Two photosynthetic antenna complexes aid in light absorption: the Fenna-Matthews-Olson complex ("FMO", also containing BChl *a*), and the chlorosomes which employ mostly BChl *c*, *d*, or *e*. Hydrogen sulfide is used as an electron source and carbon dioxide its carbon source.

Chlorobium species exhibit a dark green color; in a Winogradsky column, the green layer often observed is composed of *Chlorobium*. This genus lives in strictly anaerobic conditions below the surface of a body of water, commonly the anaerobic zone of a eutrophic lake.

Chlorobium aggregatum is a species which exists in a symbiotic relationship with a colorless, nonphotosynthetic bacteria. This species looks like a bundle of green bacteria, attached to a central rod-like cell which can move around with a flagellum. The green, outer bacteria use light to oxidize sulfide into sulfate. The inner cell, which is not able to perform photosynthesis, reduces the sulfate into sulfide. These bacteria divide in unison, giving the structure a multicellular appearance which is highly unusual in bacteria.

Chlorobium species are thought to have played an important part in mass extinction events on Earth. If the oceans turn anoxic (due to the shutdown of ocean circulation) then *Chlorobium* would be able to out compete other photosynthetic life. They would produce huge quantities of methane and hydrogen sulfide which would cause global warming and acid rain. This would have huge consequences for other oceanic organisms and also for terrestrial organisms. Evidence for abundant *Chlorobium* populations is provided by chemical fossils found in sediments deposited at the Cretaceous mass extinction.

The complete *C. tepidum* genome, which consists of 2.15 megabases (Mb), was sequenced and published in 2002. It synthesizes chlorophyll *a* and bacteriochlorophylls (BChls) *a* and *c*, of which the model organism has been used to elucidate the biosynthesis of BChl *c*. Several of its carotenoid metabolic pathways (including a novel lycopene cyclase) have similar counterparts in cyanobacteria.

Molecular Signatures for Chlorobi

Comparative genomic analysis has led to the identification of 2 conserved signature indels which are uniquely found in members of the phylum *Chlorobi* and are thus characteristic of the phylum. The first indel is a 28-amino-acid insertion in DNA polymerase III and the second is a 12 to 14 amino acid insertion in alanyl-tRNA synthetase. These indels are not found in any other bacteria and thus serve as molecular markers for the phylum. In addition to the conserved signature indels, 51 proteins which are uniquely found in members of the phylum *Chlorobi*. 65 other proteins have been identified which are unique to the *Chlorobi* phylum, however these proteins are missing in several *Chlorobi* species and are not distributed throughout the phylum with any clear pattern. This means that significant gene loss may have occurred, or the presence of these proteins may be a result of horizontal gene transfer. Of these 65 proteins, 8 are found only in *Chlorobium luteolum* and *Chlorobium phaeovibri- oides*. These two species form a strongly supported clade in phylogenetic trees and a close relationship between these species is further supported by the unique sharing of these 8 proteins.

Relatedness of Chlorobi to Bacteroidetes and Fibrobacteres Phyla

Species from the *Bacteroidetes* and *Chlorobi* phyla branch very closely together in phylogenetic trees, indicating a close relationship. Through the use of comparative genomic analysis, 3 proteins have been identified which are uniquely shared by virtually all members of the *Bacteroidetes* and *Chlorobi* phyla. The sharing of these 3 proteins is significant because other than these 3 proteins, no proteins from either the *Bacteroidetes* or *Chlorobi* phyla are shared by any other groups of bacteria. Several conserved signature indels have also been identified which are uniquely shared by members of the *Bacteroidetes* and *Chlorobi* phyla. The presence of these molecular signatures supports the close relationship of the *Bacteroidetes* and *Chlorobi* phyla. Additionally, the phylum *Fibrobacteres* is indicated to be specifically related to these two phyla. A clade consisting of these three phyla is strongly supported by phylogenetic analyses based upon a number of different proteins. These phyla also branch in the same position based upon conserved signature indels in a number of important proteins. Lastly and most importantly, two conserved signature indels (in the RpoC protein and in serine hydroxymethyltransferase) and one signature protein PG00081 have been identified that are uniquely shared by all of the species from these three phyla. All of these results provide compelling evidence that the species from these three phyla shared a common ancestor exclusive of all other bacteria and it has been proposed that they should all recognized as part of a single "FCB" superphylum.

CHLOROBIUM CHLOROCHROMATII

Chlorobium chlorochromatii, originally known as *Chlorobium aggregatum*, is a symbiotic green sulfur bacteria that performs anoxygenic photosynthesis and functions as an obligate photoautotroph using reduced sulfur species as electron donors. *Chlorobium chlorochromatii* can be found in stratified freshwater lakes.

C. chlorochromatii is a Gram-negative, non-motile bacillus, that exist in short chains. They are green in color and have a ring of chlorosomes around that line the inside of their cell wall. Within these chlorosomes contain the light harvesting pigment bacteriochlorophyll a and bacteriochlorophyll c which feed electrons into Photosystem 1.

Ecology

Photosynthetic green sulfur bacteria such as *Chlorobium chlorochromatii* reside in freshwater, stratified lakes beneath the micro-aerophilic algal layer in anaerobic, light-exposed environments. They have been found worldwide, mostly in holomictic or meromictic stratified lakes. Lakes that support this environment have been found in Germany, Tasmania, the USA, ice-covered lakes in Antarctica, Israel and Japan. *Chlorobium chlorochromatii* prefer environments with low temperature and low sulfur concentrations.

Symbiosis

Chlorobium chlorochromatii, strain CaD, was originally isolated from the phototrophic microbial consortium *Chlorochromatium aggregatum*. The ability of this epibiont to grow in pure culture indicates that it is not an obligately symbiotic organism. Despite this fact, *C. chlorochromatii* has never been found in a free-living state in naturally occurring bacterial communities.

Metabolism

C. chlorochromatii conducts anoxygenic photosynthesis which means it does not produce oxygen as a waste product like plants and cyanobacteria, this type of photosynthesis is exclusive to Bacteria. In their electron transport chain reduced forms of sulfur, e.g., H_2S. These reduced forms of sulfur are used in the electron transport chain cyclic Photosystem 1 as electron donors to reduce $NADP^+$ to NADPH. It grows under strictly anaerobic conditions as a photolithoautotroph. They grow best at a pH of 7.0-7.3 at 25 C in continuous light and absorb light maximally at 748 and 453 nm.

Genome Structure

C. chlorochromatii contains a circular genome that contains 2,572,079 bp. There are a total of 2047 genes contained on its chromosome. Of these 2047 genes, there are 1999 protein coding genes and 48 RNA coding genes. There are no pseudogenes contained within the chromosome of *C. chlorochromatii*. Data from DNA analysis suggests that genomes of green sulfur bacteria range between 2-3.3 Mb. From these data, it can be assumed that the total genome size of the consortium of *C. aggregatum* is less than 10 Mb in length.

CHLOROFLEXUS AURANTIACUS

Chloroflexus aurantiacus is a photosynthetic bacterium isolated from hot springs, belonging to the green non-sulfur bacteria. This organism is thermophilic and can grow at temperatures from 35 °C to 70 °C (94.998 to 158 °F). *Chloroflexus aurantiacus* can survive in the dark if oxygen is available. When grown in the dark, *Chloroflexus aurantiacus* has a dark orange color. When grown in sunlight it is dark green. The individual bacteria tend to form filamentous colonies enclosed in sheaths, which are known as trichomes.

Thermophilic Organisms

Physiology

As a genus, *Chloroflexus* spp. are filamentous anoxygenic phototrophic (FAP) organisms that utilize type II photosynthetic reaction centers containing bacteriochlorophyll *a* similar to the purple bacteria, and light-harvesting chlorosomes containing bacteriochlorophyll *c* similar to green sulfur bacteria of the *Chlorobi*. Like other members of its phylum, the species stains Gram negative, yet has a single lipid layer (monoderm), but with thin peptidoglycan, which is compensated for by S-layer protein.

As the name implies, these anoxygenic phototrophs do not produce oxygen as a by-product of photosynthesis, in contrast to oxygenic phototrophs such as cyanobacteria, algae, and higher plants. While oxygenic phototrophs use water as an electron donor for phototrophy, *Chloroflexus* uses reduced sulfur compounds such as hydrogen sulfide, thiosulfate, or elemental sulfur. This belies their antiquated name green non-sulfur bacteria, however *Chloroflexus* spp. can also utilize hydrogen(H_2) as a source of electrons.

Chloroflexus aurantiacus is thought to grow photoheterotrophically in nature, but it has the capability of fixing inorganic carbon through photoautotrophic growth. Instead of using the Calvin-Benson-Bassham Cycle typical of plants, *Chloroflexus aurantiacus* has been demonstrated to use an autotrophic pathway known as the 3-Hydroxypropionate pathway.

The complete electron transport chain for *Chloroflexus* spp. is not yet known. Particularly, *Chloroflexus aurantiacus* has not been demonstrated to have a cytochrome bc_1 complex, and may use different proteins to reduce cytochrome *c*.

GREEN SULFUR BACTERIA

The green sulfur bacteria (Chlorobiaceae) are a family of obligately anaerobic photoautotrophic bacteria. Together with the non-photosynthetic Ignavibacteriaceae, they form the phylum Chlorobi.

Green sulfur bacteria are nonmotile (except *Chloroherpeton thalassium*, which may glide) and capable of anoxygenic photosynthesis. In contrast to plants, green sulfur bacteria mainly use sulfide ions as electron donors. They are autotrophs that utilize the reverse tricarboxylic acid cycle to fix carbon dioxide. Green sulfur bacteria have been found in depths of up to 145m in the Black Sea, with low light availability.

Characteristics of Green-sulfur Bacteria

- Major photosynthetic pigment: Bacteriochlorophylls a plus c, d or e.

- Location of photosynthetic pigment: Chlorosomes and plasma membranes.

- Photosynthetic electron donor: H_2, H_2S, S.

- Sulfur deposition: Outside of the cell.

- Metbolic type: Photolithoautotrophs.

Metabolism

Catabolism

Photosynthesis is achieved using a Type 1 reaction centre, which contains bacteriochlorophyll *a*, and is taken place in chlorosomes. Type 1 reaction centre is equivalent to photosystem I found in plants and cyanobacteria. Green sulfur bacteria use sulfide ions, hydrogen or ferrous iron as electron donors and the process is mediated by the Type I reaction centre and Fenna-Matthews-Olson complex. Reaction centre contains bacteriochlorophylls, P840, which donates electrons to cytochrome c-551 when it is excited by light. Cytochrome c-551 then passes the electrons down the electron chain. P840 is returned to its reduced state by the oxidation of sulfide. Sulfide donates two electrons to yield elemental sulfur. Elemental sulfur is deposited in globules on the extracellular side of the outer membrane. When sulfide is depleted, the sulfur globules are consumed and oxidized to sulfate. However, the pathway of sulfur oxidation is not well-understood.

Anabolism

These autotrophs fix carbon dioxide using the reverse tricarboxylic acid (RTCA) cycle. Energy is consumed to incorporate carbon dioxide in order to assimilate pyruvate and acetate and generate macromolecules. *Chlorobium tepidum*, a member of green sulfur bacteria was found to be mixotroph due to its ability to use inorganic and organic carbon sources. They can assimilate acetate through the oxidative (forward) TCA (OTCA) cycle in addition to RTCA. In contrast to the RTCA cycle, energy is generated in the OTCA cycle, which may contribute to better growth. However, the capacity of the OTCA cycle is limited because gene that code for enzymes of the OTCA cycle are down-regulated when the bacteria is growing phototrophically.

Habitat

The Black Sea, an extremely anoxic environment, was found to house a large population of green sulfur bacteria at about 100 m depth. Due to the lack of light available in this region of the sea, most bacteria were photosynthetically inactive. The photosynthetic activity detected in the sulfide chemocline suggests that the bacteria need very little energy for cellular maintenance.

A species of green sulfur bacteria has been found living near a black smoker off the coast of Mexico at a depth of 2,500 m in the Pacific Ocean. At this depth, the bacterium, designated GSB1, lives off the dim glow of the thermal vent since no sunlight can penetrate to that depth.

Photosynthesis in the Green Sulfur Bacteria

The green sulfur bacteria use PS I for photosynthesis. Thousands of bacteriochlorophyll(BCHl) c, d and e of the cells absorb light at 720-750 nm, and the light energy is transferred to BChl a-795 and a-808 before being transferred to Fenna-Matthews-Olson (FMO)-proteins which are connected to reaction centers (RC). The FMO complex then transfers the excitation energy to the RC with its special pair which absorbs at 840 nm in the plasma membrane.

After the reaction centers receive the energy, electrons are ejected and transferred through electron transport chains (ETCs). Some electrons form Fe-S proteins in electron transport chains are accepted by ferredoxins (Fd) which can be involved in NAD(P) reduction and other metabolic reactions.

Carbon Fixation of Green Sulfur Bacteria

The reactions of reversal of the oxidative tricarboxylic acid cycle are catalyzed by four enzymes:

1. Pyruvate: Ferredoxin (Fd) oxidoreductase:

    Acetyl-CoA + CO_2 + 2Fdred + 2H+ $\rightleftharpoons$ pyruvate + CoA + 2Fdox.

2. ATP citrate lyase:

    ACL, acetyl-CoA + oxaloacetate + ADP + Pi $\rightleftharpoons$ citrate + CoA + ATP.

3. α-keto-glutarate: Ferredoxin oxidoreductase:

    Succinyl-CoA + CO_2 + 2Fdred + 2H+ $\rightleftharpoons$ α-ketoglutarate + CoA + 2Fdox.

4. Fumarare reductase:

    Succinate + acceptor $\rightleftharpoons$ fumarate + reduced acceptor.

RHODOBACTER SPHAEROIDES

Rhodobacter sphaeroides is a kind of purple bacterium; a group of bacteria that can obtain energy through photosynthesis. Its best growth conditions are anaerobic phototrophy (photoheterotrophic and photoautotrophic) and aerobic chemoheterotrophy in the absence of light. *R. sphaeroides* is also able to fix nitrogen. It is remarkably metabolically diverse, as it is able to grow heterotrophically via fermentation and aerobic and anaerobic respiration.

Rhodobacter sphaeroides has been isolated from deep lakes and stagnant waters.

Rhodobacter sphaeroides is one of the most pivotal organisms in the study of bacterial photosynthesis. It requires no unusual conditions for growth and is incredibly efficient. The regulation of its photosynthetic machinery is of great interest to researchers, as *R. sphaeroides* has an intricate system for sensing O_2 tensions. Also, when exposed to a reduction in the partial pressure of oxygen, *R. sphaeroides* develops invaginations in its cellular membrane. The photosynthetic apparatus is housed in these invaginations. These invaginations are also known as chromatophores.

The genome of *R. sphaeroides* is also somewhat intriguing. It has two chromosomes, one of 3 Mb (CI) and one of 900 Kb (CII), and five naturally occurring plasmids. Many genes are duplicated between the two chromosomes but appear to be differentially regulated. Moreover, many of the open reading frames (ORFs) on CII seem to code for proteins of unknown function. When genes of unknown function on CII are disrupted, many types of auxotrophy result, emphasizing that the CII is not merely a truncated version of CI.

Small Non-coding RNA

Bacterial small RNAs have been identified as components of many regulatory networks. Twenty sRNAs were experimentally identified in *Rhodobacter spheroids,* and the abundant ones were shown to be affected by singlet oxygen ($^1O_2$) exposure. $^1O_2$ which generates photooxidative stress, is made by bacteriochlorophyll upon exposure to oxygen and light. One of the $^1O_2$ induced sRNAs SorY ($^1O_2$ resistance RNA Y) was shown to be induced under several stress conditions and conferred resistance against $^1O_2$ by affecting a metabolite transporter. SorX is the second $^1O_2$ induced sRNA that counteracts oxidative stress by targeting mRNA for a transporter. It also has an impact on resistance against organic hydroperoxides. A cluster of four homologous sRNAs called CcsR for conserved CCUCCUCCC motif stress-induced RNA has been shown to play a role in photo-oxidative stress resistance as well. PcrZ (photosynthesis control RNA Z) identified in *R. sphaeroides,* is a *trans*-acting sRNA which counteracts the redox-dependent induction of photosynthesis genes, mediated by protein regulators.

PURPLE BACTERIA

Purple bacteria grown in winogradsky column.

Purple bacteria or purple photosynthetic bacteria are proteobacteria that are phototrophic, that is, capable of producing their own food via photosynthesis. They are pigmented with bacteriochlorophyll *a* or *b*, together with various carotenoids, which give them colours ranging between purple, red, brown, and orange. They may be divided into two groups – purple sulfur bacteria (Chromatiales, in part) and purple non-sulfur bacteria (Rhodospirillaceae).

Metabolism

Purple bacteria are mainly photoautotrophic, but are also known to be chemoautotrophic and photoheterotrophic. They can be mixotrophs, capable of aerobic respiration and fermentation.

Location

Photosynthesis occurs at reaction centers on the cell membrane, where the photosynthetic pigments (i.e. bacteriochlorophyll, carotenoids) and pigment-binding proteins are invaginated to form vesicle sacs, tubules, or single-paired or stacked lamellar sheets. This is called the intracytoplasmic membrane (ICM) which has increased surface area to maximize light absorption.

The Purple non-sulfur bacteria *Rhodospirillum*.

Mechanism

Purple bacteria use cyclic electron transport driven by a series of redox reactions. Light-harvesting complexes surrounding a reaction centre (RC) harvest photons in the form of resonance energy, exciting chlorophyll pigments P870 or P960 located in the RC. Excited electrons are cycled from P870 to quinones Q_A and Q_B, then passed to cytochrome bc_1, cytochrome c_2, and back to P870. The reduced quinone Q_B attracts two cytoplasmic protons and becomes QH_2, eventually being oxidized and releasing the protons to be pumped into the periplasm by the cytochrome bc_1 complex. The resulting charge separation between the cytoplasm and periplasm generates a proton motive force used by ATP synthase to produce ATP energy.

Electron Donors for Anabolism

Purple bacteria also transfer electrons from external electron donors directly to cytochrome bc_1 to generate NADH or NADPH used for anabolism. They are anoxygenic because they do not use water as an electron donor to produce oxygen. One type of purple bacteria, called purple sulfur bacteria (PSB), use sulfide or sulfur as electron donors. Another type, called purple non-sulfur bacteria, typically use hydrogen as an electron donor but can also use sulfide or organic compounds at lower concentrations compared to PSB.

Purple bacteria lack external electron carriers to spontaneously reduce $NAD(P)^+$ to $NAD(P)H$, so they must use their reduced quinones to endergonically reduce $NAD(P)^+$. This process is driven by the proton motive force and is called reverse electron flow.

CYANOBACTERIA

Cyanobacteria, also referred to as blue-green algae and blue-green bacteria is a bacteria phylum that obtain their energy through a process known as photosynthesis. Because they require the basic environmental conditions, this bacteria can be found in a variety of environments ranging from marine to terrestrial habitats.

Cyanobacteria is also composed of a wide variety of bacteria species of different shapes are sizes that can be found in different habitats in the environment. These are spread across the 150 genera that have been identified so far and play various important roles in nature.

Examples include:

- Microcystis aeruginosa.

- Cylindrospermopsis raciborskii.

- Anabaena circinalis.

- Cyanophora paradoxa.

- Nostoc commune.

Taxonomy/Classification

In 1985, the proposed classification of cyanobacteria took into account the Bacteriological factor. The proposal identified four Orders of the bacteria which included Chroococcales, Nostocales, Oscillatoriales and Stigonematales. However, other orders of the phylum that have been discovered include Chroococcales, Gloeobacterales, and Pleurocapsales.

The bacteria also falls under Kingdom Monera and Division Eubacteria. Further classification has however resulted in significant debate at higher taxonomic levels.

Initially, they were classified as blue-green algae because they possess chlorophyll and algal-like appearance. However, further studies showed that they are prokaryotic, which helped re-classify them appropriately.

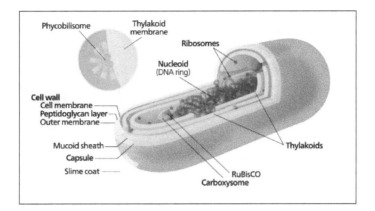

Characteristics based on their Order

Its structural diversity has been shown to be considerable. For this reason, species had to be grouped into categories that have similar characteristics. This section will focus on the major Orders under the phylum to highlight their respective traits.

Chroococcales

This Order is composed of two main classes (Chroococcaceae and Entophysalidaceae). Species in these classes are coccoid or rod-shaped (with Entophysalidaceae being largely composed of coccoid organisms).

Some of the other characteristics among these organisms include the fact that they reproduce through binary fission and they can create colonies to form dense masses that can be seen on such surfaces as moist rocks.

Members of Order Chroococcales also have the following characteristics:

- A spherical, ovoid or cylindrical cell shape structure.

- They may occur singly following cell division.

- As they mature, they aggregate to form colonies that are held together by a slimy matrix.

Other members of Order Chroococcales include:

- Pleurocapsa

- Aphanocapsa

- Merismopedia

- Gloeocapsa

- Microcystis

Pleurocapsales

Some of the unifying characteristics of Order Pleurocapsales are that they all reproduce through multiple fission in addition to releasing endospores. Compared to other organisms that also form endospores, members of Pleurocapsales divide through binary and multiple fission, which is the main distinguishing factor.

Here, the enlargement of the spores is followed by additional binary fission to produce a mass of vegetative cells. Some of the cells in the mass go through multiple fission to release more endospores.

The order is also composed of a wide range of organisms that can be found in an array of habitats ranging from terrestrial and marine environments. As they develop, some members of this order have been shown to develop as epiphytes on algae and as epiliths on such surfaces as moist rocks.

They can also produce pseudo-filaments that can reproduce through baeocytes.

Members of Order Pleurocapsales include:

- Chroococcidiopsis

- Pleurocapsa

- Dermocarpella

- Xenococcus

Oscillatoriales

The Order Oscillatoriales is largely composed of filamentous cyanobacteria (having uniseriate filaments). However, members of this group lack true branching, akinetes and heterocysts.

Members of this Order can be found in a variety of environments from fresh water and saline water bodies to terrestrial habitats.

Some of the other traits of this Order include:

- They tend to form multicellular elongated structures.

- They are trichal/filamentous.

Members of Oscillatoriales include:

- Phormidium

- Microcoleus

- Lyngbya

- Planktothrix

Nostocales

Like Stigonematales, members of Order Nostocales have heterogenus cellular composition in their trichomes. The vegetative cells in this order are also divided into heterocysts that have a thick hyaline protoplast and are involved in nitrogen fixation as well as akinetes that have in place thick cell walls that allow them to survive when conditions are unfavourable.

Some of the other traits include:

- False branching in some species.

Some members of this Order include:

- Cylindrospermopsis

- Calothrix

- Anabaena

- Nostoc

Stigonematales

Members of Order Stigonematales share several similar traits to those of Order Nostocales. These include such traits as trichomes with heterogeneous cellular composition as well as heterocysts and akinetes vegetative cells.

Some of the other traits associated with the order Stigonematales include multiseriated filaments with true branching.

Members of this Order include:

- Stigonema

- Mastigocladus

Gloebacterales

Consisting of such organisms such as the members of Gloeobacter, Gloeobacterales possess phycobilisomes which are light harvesting complexes. As a result, Gloeobacterales like Gloeobacter lack thylakoids that are found in other members of cyanobacteria. However, like other cyanobacteria, they can be found in a range of environments like limestone rocks and other aquatic environments.

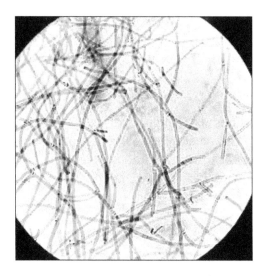

Occurrence of Cyanobacteria in Nature

While the Phylum cyanobacteria is composed of diverse members, there are a number of traits that can be found in the majority, if not all cyanobacteria species.

For instance, a majority of the species are aerobicphotoautotrophs. As such, they are heavily reliant on water, carbon dioxide, light as well as various inorganic substances for their life processes.

Given that photosynthesis is the primary mode of energy production among these organisms, they are heavily dependent on sunlight, carbon dioxide and water for this mechanism to be successful.

For their survival, a good number have been shown to be able to adapt through a number of mechanisms. A good example of this includes ultraviolet absorbing sheath pigments that allows some of the species to be able to survive exposed environments.

They are particularly prominent in cold/hot springs; marine bodies as well as others areas where microalgae are incapable of surviving. This is made possible by the fact that they are capable of adapting to different environments.

Cyanobacteria play an important role within the microbial community; for this reason, they can be found almost everywhere on our planet. However, some of these environments are unfavourable which makes it difficult for many other live organisms to survive in them. These include such environments as deserts, very hot springs and Antarctic ice shelves among others.

For this reason, they need to adapt in various ways in order to survive these conditions. In addition to the ultraviolet absorbing sheath pigments found in some species, many planktonic cyanobacteria have been shown to be able to survive Ultra Violet Rays (UVR) by either floating or sinking in their environment by using gas vacuoles.

For others like Oscillatoriales, survival is enhanced by actively moving from microbial mat surfaces into the matrix sediments which provide more favorable conditions. This allows the organisms to survive instances of UVR. In Antarctic Lakes, Cyanobacteria have been shown to significantly lower their oxygen consumption.

The presence of chlorophyll is one of the characteristics identified. However, it is worth noting that they also have other accessory pigments that include allophycocyanin, phycoerythrin and phycocyanin. These pigments play an important role given that they help effectively utilize the light spectrum in a given region/environment.

In the event of varying light spectrums, the organisms can still carry on their photosynthetic process which in turn enhance their survival. In addition, members of cyanobacteria are capable of storing various important nutrients and metabolites within their cytoplasm.

This, in addition to dinitrogen fixations, provides them with the necessary nutrition they require to survive. Among other mechanisms, this helps the bacteria survive in deep and dark environments for a long period of time. Through such adaptations, members have been able to survive in different environments, both favorable and extreme, on earth.

Some of the other survival strategies used include:

- The production of mucopolysaccharides - this slows down liquid flow during freeze and thaw.

- Production of compatible solutes (sucrose, glucosylglycerol etc) - produced by water-stressed organisms.

- The ability to fix carbon dioxide at very low water potentials.

Symbioses

Because of their ability to fix nitrogen, heterocystous cyanobacteria develop symbiotic relationships with a variety of eukaryotic plant species such as algae, liverworts, angiospores and ferns among others.

As studies have shown, development of symbiotic relationships between free-living cyanobacteria and other eukaryotic plants results in morphological, physiological and biochemical modification thus causing a change to the original morphology, physiology and biochemical form of the bacteria.

During nitrogen fixation, an enzyme known as nitrogenase transfers several pairs (three pairs) of electrons and protons to the nitrogen molecule to create two ammonia molecules. This is then added to glutamic acid to form glutamine (in the plant's cells) that in turn provides amide nitrogen used for the synthesis of amino acids as well as components of DNA and RNA. In this relationship, cyanobacteria, which are nitrogen-fixers benefit from carbon dioxide that is produced by the host (algae etc).

With fungi, cyanobacteria symbiosis produces an association known as lichens. The symbiotic relationship (lichenization) between fungi and cyanobacteria only accounts for 10 percent of all lichens.

Given that fungi are incapable of photosynthesis; this relationship is beneficial to the fungi given that they can receive various nutrients and thus energy from the photosynthetic activities of the cyanobacteria.

Gram Negative

Cyanobacteria have a gram-negative structure, which means that they cannot retain the primary stain duringgram-staining. However, microscopic observation of the species has shown that they have a thicker peptidoglycan layer compared to other gram-negative bacteria.

Studies on the peptidoglycan layer has also revealed a more complex structure (with the cross-link between the peptidoglycan chains being higher compared to other organisms).

Toxins and Treatment

In their environment, cyanobacteria produce a wide range of compounds such as 2-methylisoborneol and geosmin. When released in water, these compounds can result in a change in the taste and odor.

Apart from these compounds, they have also been shown to produce such toxins as hepatotoxins, neurotoxins and dermatoxins. These toxins, often released by planktonic species can have negative impacts on human beings in causing infections of the liver (hepatotoxins), the nervous system (neurotoxins) and skin irritation which is caused by dermatotoxins.

Cyanobacteria are also responsible for microcystins (cyclic peptides that may either contain seven amino acids or other constituents that resemble amino-acid). These compounds are synthesized by enzyme complexes. The toxin produced through this process can cause serious health issues and death of metazoa.

Because of these toxins, the presence of cyanobacteria in drinking reservoirs is of great concern since they can have serious health impacts on all animals including human beings.

Whereas some of the toxins are released in water, some of these bacteria have been shown to release potent toxins such as anatoxin-a which is produced by Anabaena flosaquae that can cause death within 30 minutes.

Some of the toxins (capable of causing paralysis e.g. neosaxitoxin produced by Aphanizornenon flosaquae) have been identified in some blooms and can also result in death.

Following contact with algal blooms, these toxins can cause allergic rhinitis and dermatitis among other symptoms. Bloom forming species are often found in environments with favorable conditions (warm, stable with high nutrient levels). Some of the organisms identified here include members of Anabaena, Microcystis and Aphanizomenon.

Some of the signs/symptoms in human beings include (impacts of the peptide toxins):

- Diarrhea

- Pilo-erection

- Vomiting

- General weakness

- Cold

- Death

Effective treatment is yet to be developed and thus prevention is the best means of avoiding such infections. Here, prevention involves not drinking water from lakes and rivers (particularly those with blooms and scum).

Such water should also not be used for such purposes as cooking or preparing refreshments given that they may contain the toxins. While some may think of boiling such water for safe consumption, this is also not recommended because doing so may increase the effects of the toxic substances.

Depending on the infection and the type of cyanobacteria, some of the methods used in treatment include:

- Intravenous electrolytes.

- Oxygen therapy.

- Antibiotics.

- Vitamins.

- Use of drugs such as cholestyramine.

Permissions

All chapters in this book are published with permission under the Creative Commons Attribution Share Alike License or equivalent. Every chapter published in this book has been scrutinized by our experts. Their significance has been extensively debated. The topics covered herein carry significant information for a comprehensive understanding. They may even be implemented as practical applications or may be referred to as a beginning point for further studies.

We would like to thank the editorial team for lending their expertise to make the book truly unique. They have played a crucial role in the development of this book. Without their invaluable contributions this book wouldn't have been possible. They have made vital efforts to compile up to date information on the varied aspects of this subject to make this book a valuable addition to the collection of many professionals and students.

This book was conceptualized with the vision of imparting up-to-date and integrated information in this field. To ensure the same, a matchless editorial board was set up. Every individual on the board went through rigorous rounds of assessment to prove their worth. After which they invested a large part of their time researching and compiling the most relevant data for our readers.

The editorial board has been involved in producing this book since its inception. They have spent rigorous hours researching and exploring the diverse topics which have resulted in the successful publishing of this book. They have passed on their knowledge of decades through this book. To expedite this challenging task, the publisher supported the team at every step. A small team of assistant editors was also appointed to further simplify the editing procedure and attain best results for the readers.

Apart from the editorial board, the designing team has also invested a significant amount of their time in understanding the subject and creating the most relevant covers. They scrutinized every image to scout for the most suitable representation of the subject and create an appropriate cover for the book.

The publishing team has been an ardent support to the editorial, designing and production team. Their endless efforts to recruit the best for this project, has resulted in the accomplishment of this book. They are a veteran in the field of academics and their pool of knowledge is as vast as their experience in printing. Their expertise and guidance has proved useful at every step. Their uncompromising quality standards have made this book an exceptional effort. Their encouragement from time to time has been an inspiration for everyone.

The publisher and the editorial board hope that this book will prove to be a valuable piece of knowledge for students, practitioners and scholars across the globe.

Index

CPSIA information can be obtained
at www.ICGtesting.com
Printed in the USA
BVHW010145270322
632559BV00003B/77